T0135438

Semi-Explicit MPC for Classes of Linear and Nonlinear Systems

Von der Fakultät Konstruktions-, Produktions- und Fahrzeugtechnik
der Universität Stuttgart zur Erlangung der Würde eines
Doktor-Ingenieurs (Dr.-Ing.) genehmigte Abhandlung

Vorgelegt von

Gregor Goebel

aus Reutlingen

Hauptberichter: Prof. Dr.-Ing. Frank Allgöwer
Mitberichter: Prof. Dr.-Ing. Martin Mönnigmann
Asst. Prof. Colin Jones, PhD

Tag der mündlichen Prüfung: 24. April 2018

Institut für Systemtheorie und Regelungstechnik

Universität Stuttgart

2018

Bibliografische Information der Deutschen Nationalbibliothek

Die Deutsche Nationalbibliothek verzeichnet diese Publikation in der
Deutschen Nationalbibliografie; detaillierte bibliografische Daten sind
im Internet über http://dnb.d-nb.de abrufbar.

D 93

ISBN 978-3-8325-4884-1

Logos Verlag Berlin GmbH
Comeniushof, Gubener Str. 47,
10243 Berlin
Tel.: +49 (0)30 42 85 10 90
Fax: +49 (0)30 42 85 10 92
INTERNET: https://www.logos-verlag.de

Acknowledgements

The results of this thesis were developed during my employment as a research and teaching assistant at the Institute for Systems Theory and Automatic Control (IST) at the University of Stuttgart. Many people accompanied and supported me during this period and contributed towards making this time fruitful and pleasant for me at the same time. I want to express my deep gratitude to these people.

First and foremost, I am indebted to my advisor Professor Frank Allgöwer. Right from the beginning of my Engineering Cybernetics studies, he inspired me very much – not only as a control engineer and teacher but also with his personality. During my time as a PhD student at his institute, he provided me outstanding opportunities, extensive scientific freedom and extremely inspiring working conditions. The trust he had in my work was very encouraging for me. I thank Professor Allgöwer very much for all this.

I am grateful to Professor Martin Mönnigmann, Professor Colin Jones and Professor Oliver Riedel for their interest in my work, for being members of my doctoral examination committee and for the very pleasant atmosphere they created during the examination itself.

I especially thank my former colleagues and fellow PhD students at the IST of which many became friends. Their vital contributions ranged from scientific discussions and proof-reading to pleasant distractions at the IST when appropriate to free-time activities far beyond anything scientific or PhD related.

In fact, I am grateful to numerous other people such as friends, students and scientists which I met during my time at the IST. I am sorry that I cannot name all of them here.

Finally, I wholeheartedly thank my family and especially my partner Verena for their continuous support. They were always there for me, they encouraged and inspired me. I feel privileged to have them in my life.

Stuttgart, April 2018
Gregor Goebel

Table of Contents

Abstract

In this thesis, we propose and elaborate a novel fast and efficient model predictive control (MPC) scheme. The development of such schemes is a recent and ongoing scientific endeavor in order to enable new applications of the general MPC paradigm. To this end, so far mainly two separate routes have been followed. A first branch simplifies the online numerical solution of the MPC optimization problem during runtime, whereas a second branch, termed *explicit MPC*, pre-computes offline the solution of the optimization problem as a function of the state and online only evaluates this function. The MPC scheme proposed here combines aspects of both paradigms in order to reduce their individual downsides and combine their benefits. In more detail, we propose a novel, generally applicable *semi-explicit MPC* approach which consists of an offline and an online part. During the offline part, state-dependent parametrizations are computed data based. Applying a tailored subspace clustering algorithm, they are determined such that they optimally approximate solutions of the MPC optimization problem and exploit state-dependence of the solutions for this purpose. During the subsequent online part, the parametrizations are employed in the optimization such that overall the online MPC scheme is simplified considerably. Adjusting the complexity of the parametrizations allows to trade off the amount of used explicit information against the simplicity of the remaining online optimization.

We first propose our general approach and evaluate its general properties. Then, specific algorithms and results are developed for linear systems and for two classes of nonlinear systems. We contribute different approaches to compute parametrizations which are guaranteed to result in a feasible parametrized optimization for a given set of states. These approaches are applicable to different types of problems and each have different strengths. We give guarantees on system theoretic properties of the semi-explicit MPC scheme including closed-loop asymptotic stability, closed-loop control performance and recursive feasibility. Furthermore, we formulate and evaluate an extremely simple numerical optimization procedure which is applicable to linear problems in combination with univariate parametrizations of the proposed type.

Overall, readily applicable semi-explicit MPC algorithms for different classes of systems are formulated and a corresponding theoretic foundation is provided. We evaluate the results in several numerical examples. These examples verify that the results presented in this thesis are not only an interesting conceptual step forward but, in fact, can have advantages over existing MPC schemes and can be applicable beyond them.

Deutsche Kurzfassung

Semi-explizites MPC für Klassen von linearen und nichtlinearen Systemen

Motivation und Problemstellung

Die vorliegende Arbeit befasst sich mit der Entwicklung effizienter Algorithmen zur modell-prädiktiven Regelung. Unter dem Begriff (modell-)prädiktive Regelung, kurz MPC (von engl. *model predictive control*), wird eine Klasse von modernen Regelungsalgorithmen verstanden, in denen der Stelleingang mittels der wiederholten Lösung eines Optimalsteuerungsproblems mit endlichem Zeithorizont berechnet wird. Dabei wird typischerweise ein entsprechendes Optimierungsproblem zur Laufzeit numerisch gelöst, wobei jeweils der aktuelle System-zustand als Anfangsbedingung verwendet wird und lediglich der erste Teil der optimalen Eingangssequenz angewendet wird, bevor das Prozedere mit neuer Anfangsbedingung und über einen verschobenen Zeithorizont wiederholt wird. Aus diesem Prinzip ergeben sich drei wesentliche Vorteile, welche MPC gegenüber anderen klassischen Regelungsverfahren auszeichnet: Bestehende Regelziele können im Gütekriterium des Optimalsteuerungspro-blems formuliert und somit direkt verfolgt werden, Beschränkungen der Systemeingänge und der Systemzustände können explizit berücksichtigt werden und die Herangehensweise ist für nichtlineare Systeme ebenso wie für Systeme mit mehreren Ein- und Ausgängen anwendbar. Diese Vorteile haben dazu geführt, dass MPC heute nicht nur Gegenstand intensiver theoretischer Forschungen ist, sondern auch in diversen praktischen Anwendungen erfolgreich eingesetzt wird, siehe (Lee, 2011; Mayne et al., 2000; Qin and Badgwell, 2003).

Ein wesentlicher Nachteil des MPC-Ansatzes ergibt sich aus dem erforderlichen Re-chenaufwand, um das Optimalsteuerungsproblem numerisch zu lösen. Dies schränkt die Anwendbarkeit von MPC auf Systeme ein, deren Zeitkonstanten relativ geringe Abtastraten und damit ausreichende Rechenzeiten zulassen. Klassische Einsatzgebiete der prädiktiven Regelung liegen daher vor allem im Bereich der verfahrenstechnischen Anlagen. Einerseits verfügen die Regelstrecken hier über relativ große Zeitkonstanten, andererseits können in die-sen Anwendungen leistungsstarke dedizierte Rechner zur Ausführung der Regelalgorithmen eingesetzt werden, (Qin and Badgwell, 2003).

Aktuelle Bestrebungen zielen jedoch mehr und mehr darauf ab, MPC und seine Vorteile auch nutzbar zu machen für schnellere Anwendungen und solche Anwendungen, in denen nur leistungsschwächere Rechenhardware verfügbar ist. Typische neue und zukünftig angestrebte Einsatzgebiete der prädiktiven Regelung umfassen dabei mechatronische Anwendungen sowie insbesondere Anwendungen aus dem Automobilbereich (Del Re et al., 2010; Di Cairano, 2012; Hrovat et al., 2012). Besonders attraktiv am Einsatz von MPC in diesen Bereichen sind die bezüglich des geregelten Systems verfügbaren Garantien und das durch MPC optimierte Systemverhalten. Das Ziel, derartige schnelle Systeme mittels MPC bei gleichzeitigem

Einsatz von kostengünstiger und daher leistungsschwacher Rechenhardware zu regeln, erfordert spezielle und besonders effiziente Algorithmen. Solche Algorithmen müssen äußerst geringe Rechenzeiten ermöglichen und sich dabei gleichzeitig extrem ressourceneffizient verhalten.

Die Entwicklung derartiger Algorithmen ist ein wesentlicher Gegenstand aktueller Forschung im Bereich der prädiktiven Regelung. Aktuelle Bestrebungen zielen einerseits darauf ab, die Lösung des Optimierungsproblems zur Laufzeit zu vereinfachen. Hierbei zum Einsatz kommen beispielsweise Parametrisierungen der Entscheidungsvariablen im Optimierungsproblem, siehe unter anderem (Alamir, 2012; Cagienard et al., 2007; Shekhar and Manzie, 2015), sowie besonders maßgeschneiderte Optimierungsalgorithmen, siehe beispielsweise (Domahidi et al., 2012; Jerez et al., 2014; Wang and Boyd, 2010). Dabei maßgeblich ungenutzt bleibt jedoch die Tatsache, dass es sich beim eingesetzten Optimalsteuerungsproblem tatsächlich um ein zustandsabhängiges Optimierungsproblem handelt, welches implizit ein Rückführgesetz definiert. Andererseits stellt das sogenannte *explizite MPC* eine wichtige Forschungsrichtung dar, welche im Gegensatz dazu komplett auf diesem Prinzip fußt. Die Idee hierbei ist es, eine explizite Repräsentation des Rückführgesetzes vorab zu bestimmen, welche zur Laufzeit des Systems lediglich ausgewertet werden muss, siehe (Bemporad et al., 2002; Darup and Mönnigmann, 2012; Summers et al., 2010; Tøndel et al., 2003). Wenngleich dieser Ansatz eine Reihe von Vorteilen gegenüber klassischen MPC-Algorithmen besitzt, so schränkt dessen ungünstige Skalierbarkeit seine praktische Anwendbarkeit doch erheblich ein. Ursache hierfür ist, dass die Komplexität des Rückführgesetzes mit wachsender Größe des Problems sehr schnell zunimmt, sodass die Auswertung eines solchen explizit verfügbaren Rückführgesetzes mehr Ressourcen und Rechenzeit in Anspruch nimmt, als die numerische Lösung des zugrundeliegenden Optimierungsproblems.

Die komplementäre Natur der beiden genannten Ansätze ist offensichtlich, sie bleibt jedoch in aktuellen MPC-Algorithmen weitgehend ungenutzt. Es ist direkt einleuchtend, dass die Zustandsabhängigkeit des Optimierungsproblems nicht nur zu dessen expliziter Lösung genutzt werden kann, sondern ebenso das Potential zu dessen Vereinfachung hat. Dass dadurch die Skalierbarkeit gegenüber komplett expliziten Ansätzen verbessert wird, ist erwartbar. Umgekehrt ist allgemein nicht davon auszugehen, dass Algorithmen mit minimaler Komplexität die numerische Lösung des Optimierungsproblems entweder komplett vorab oder komplett zur Laufzeit vornehmen. Stattdessen können ebenso gut dazwischenliegende Varianten, also Varianten, welche nur Teile der Information des Rückführgesetzes nutzen, die geringstmögliche Komplexität besitzen. Darüber hinaus ist ein Ansatz, welcher beide Konzepte kombiniert, aus konzeptioneller Sicht sehr interessant.

Die vorliegende Arbeit widmet sich daher dem Ziel, schnelle und recheneffiziente MPC-Algorithmen zu formulieren, welche die komplementäre Beschaffenheit bestehender Ansätze ausnutzen indem sie diese kombinieren. Damit soll ein Beitrag geleistet werden zur Erweiterung der Anwendungsmöglichkeiten von prädiktiven Regelungsalgorithmen und gleichzeitig ein konzeptionell interessanter und neuartiger Ansatz vorgeschlagen und ausgearbeitet werden.

Semi-explizite Prädiktive Regelung – Der Ansatz

In der vorliegenden Arbeit stellen wir einen neuartigen *semi-expliziten MPC-Ansatz* vor, der Aspekte des klassischen und des expliziten MPCs vereint und dadurch deren Vorteile kombiniert. Die Grundidee dieses Ansatzes ist es, zustandsabhängige Parametrisierungen für die Entscheidungsvariable des Optimalsteuerungsproblems im Vorhinein zu berechnen, sodass diese zur Laufzeit des Systems zur Vereinfachung der numerischen Optimierung eingesetzt werden können. Dabei wird ein Teil des Informationsgehaltes des MPC-Rückführgesetzes in Form der zustandsabhängigen Parametrisierungen gespeichert und zur Laufzeit nutzbar gemacht. Über die Komplexität der Parametrisierungen kann dabei justiert werden, zu welchem Grad der Algorithmus explizite Informationen über das Rückführgesetz nutzt, beziehungsweise zu welchem Grad diese Information durch eine komplexere numerische Optimierung zur Laufzeit erzeugt wird. Ein zusätzlicher innovativer Aspekt ist dabei, dass die Parametrisierungen datenbasiert berechnet werden, mittels eines Clustering-Algorithmus, der auf optimale Eingangssequenzen angewendet wird. In der beschriebenen Kombination unterschiedlicher Aspekte ist die vorgeschlagene Herangehensweise einzigartig und stellt somit einen komplett neuartigen, innovativen Ansatz für effiziente MPC-Algorithmen dar, der speziell auch für nichtlineare Systeme angewendet werden kann.

Forschungsbeiträge und Gliederung der Arbeit

In der Arbeit präsentieren wir einerseits unterschiedliche Algorithmen, um für verschiedene Systemklassen Parametrisierungen der vorgeschlagenen Art zu bestimmen und hinsichtlich bestimmter Eigenschaften auszuwerten. Andererseits präsentieren wir unterschiedliche semi-explizite MPC-Algorithmen, welche die Parametrisierungen nutzen. Insgesamt werden so direkt anwendbare semi-explizite MPC-Schemata für mehrere Systemklassen vorgestellt.

Die Gewährleistung günstiger Eigenschaften der prädiktiven Regelungsalgorithmen trotz der Verwendung von Parametrisierungen ist dabei eine Herausforderung und gleichzeitig wichtige Voraussetzung zur sinnvollen praktischen Anwendbarkeit der Ergebnisse. Im Rahmen dieser Arbeit geben wir unter anderem Garantien für die vorgeschlagenen semi-expliziten Algorithmen hinsichtlich asymptotischer Stabilität und Regelgüte des geschlossenen Kreises, zur Einhaltung von Zustands- und Eingangsbeschränkungen und für die Lösbarkeit des parametrisierten Optimierungsproblems.

Im Folgenden wird ein Überblick über die Gliederung dieser Arbeit gegeben und wesentliche Beiträge der einzelnen Kapitel werden hervorgehoben.

Kapitel 2 – Grundlagen zu MPC und parametrischer Optimierung

In diesem Kapitel werden Grundlagen der präsentierten Ergebnisse mit Fokus auf die Themen stabilisierendes MPC und multiparametrische Optimierung gelegt.

Kapitel 3 – Allgemeine Ergebnisse zu semi-explizitem MPC basierend auf Unterraum-Clustering

In diesem Kapitel stellen wir unseren semi-expliziten MPC-Ansatz vor und führen die grundlegenden Versionen der dafür erforderlichen Algorithmen ein. Wir untersuchen allgemeine Eigenschaften der Algorithmen und des semi-expliziten MPC-Schemas. Dieses Kapitel basiert in Teilen auf (Goebel and Allgöwer, 2013, 2014a,b, 2017a,b). Unsere wesentlichen Forschungsbeiträge in diesem Kapitel sind:

- Wir führen die verwendeten zustandsabhängigen Parametrisierungen ein.

- Wir zeigen, dass die vorgeschlagenen Parametrisierungen sinnvoll und gewinnbringend in der parametrischen Optimierung eingesetzt werden können.

- Wir stellen einen allgemeinen Ansatz vor, um Parametrisierungen der vorgeschlagenen Art zu berechnen.

- Wir formulieren auf Basis der Parametrisierungen ein einfaches, numerisch effizientes semi-explizites MPC-Schema für allgemeine zeitdiskrete Systeme.

- Wir beweisen unterschiedliche systemtheoretische Eigenschaften des semi-expliziten MPC-Ansatzes.

Kapitel 4 – Semi-explizites MPC für lineare Systeme

In diesem Kapitel arbeiten wir den zuvor allgemein eingeführten Ansatz detailliert für die Klasse der linearen Systeme mit affinen Beschränkungen aus. Unter Ausnutzung der Eigenschaften dieser Systemklasse präsentieren wir konkrete direkt anwendbare Algorithmen für semi-explizites MPC. Dieses Kapitel basiert teilweise auf (Goebel and Allgöwer, 2013, 2014a, 2017a,b). Unsere wesentlichen Forschungsbeiträge in diesem Kapitel sind:

- Wir schlagen drei unterschiedliche Methoden zur Berechnung von Parametrisierungen vor, die garantieren, dass die parametrisierte Optimierung lösbar ist für eine bestimmte vorgegebene Menge von Zuständen.

- Wir zeigen eine obere Grenze an die Zunahme der Kosten entlang von Trajektorien des geschlossenen Kreises, welche durch die Anwendung des semi-expliziten Ansatzes verursacht wird.

- Wir formulieren einen Stabilitätstest, der genutzt werden kann, um zu verifizieren, ob eine gegebene Parametrisierung in einem vereinfachten semi-expliziten MPC-Schema genutzt werden kann und dennoch zu einem asymptotisch stabilen geschlossenen Kreis führt.

Kapitel 5 – Nahezu explizites MPC für lineare Systeme

In diesem Kapitel untersuchen wir detailliert, wie die numerische Optimierung vereinfacht werden kann, indem im Fall linearer Systeme Parametrisierungen eingesetzt werden, die in skalaren Entscheidungsvariablen resultieren. Dieses Kapitel basiert teilweise auf (Goebel and Allgöwer, 2015, 2017a). Unsere wesentlichen Forschungsbeiträge in diesem Kapitel sind:

- Wir schlagen ein extrem einfaches numerisches Optimierungsverfahren vor, das im Fall univariater Parametrisierungen eingesetzt werden kann.

- Wir vereinfachen das letztgenannte Optimierungsverfahren weiter, indem nicht benötigte Beschränkungen vorab identifiziert und beseitigt werden, sowie durch Ausnutzung einer speziellen Eigenschaft der Parametrisierung.

- Wir evaluieren die numerische Komplexität des vorgeschlagenen Optimierungsverfahrens und zeigen dessen Vorteile gegenüber dem Einsatz von expliziten Lösungen.

Kapitel 6 – Semi-explizites MPC für nichtlineare Systeme

In diesem Kapitel wird der allgemeine semi-explizite MPC-Ansatz weiter ausgearbeitet für die Anwendung für nichtlineare Systeme und spezielle Ergebnisse für zwei Klassen nichtlinearer Systeme werden präsentiert. Dieses Kapitel basiert teilweise auf (Goebel and Allgöwer, 2014b). Unsere wesentlichen Forschungsbeiträge in diesem Kapitel sind:

- Wir präsentieren einen angepassten Clustering-Algorithmus, der es ermöglicht, die Form von Clustern im Zustandsraum positiv zu beeinflussen, um dadurch die Anwendbarkeit des gesamten Ansatzes für nichtlineare Systeme zu verbessern.

- Wir stellen einen Test vor, der es ermöglicht, die Lösbarkeit eines parametrisierten nichtlinearen MPC-Optimierungsproblems für Mengen von Zuständen zu evaluieren.

- Wir formulieren einen Algorithmus, um Parametrisierungen so zu berechnen, dass die parametrisierte Optimierung garantiert lösbar ist für eine vorgegebene Menge von Zuständen.

- Die beiden letztgenannten zunächst allgemein formulierten Ergebnisse werden für die Klassen der Lipschitz-stetigen Systeme und der sogenannten inkrementell stabilen Systeme konkretisiert.

Kapitel 7 – Fazit

Schließlich fassen wir in diesem Kapitel die Ergebnisse der vorliegenden Arbeit zusammen und zeigen Richtungen für mögliche zukünftige, an diese Arbeit anknüpfende Forschung auf.

Chapter 1

Introduction

1.1 Motivation

Model predictive control (MPC) is a very successful modern control strategy which combines three striking advantages: First, constraints on system states and inputs can explicitly be incorporated. Second, a given performance criterion formulated regarding closed-loop behavior is directly optimized. Third, it is immediately applicable to nonlinear systems as well as to systems with multiple inputs. These advantages directly result from the general principle of model predictive control algorithms which allows to regard these aspects. In MPC typically an open-loop optimal control problem is solved online over a finite (prediction) time horizon using the current system state as initial condition. Only the first part of the optimal input sequence is applied to the plant and the procedure is repeated using a shifted time horizon and the updated system state as initial condition. This way, feedback is introduced. At the downside of MPC is the numerical effort needed to solve the optimal control problem during run time of the controlled system. In particular, the requirement to obtain a solution of the optimization problem sufficiently quickly with respect to the time-constants of the system to be controlled has largely restricted practical applications of MPC in the past.

Practical applications of MPC started to emerge in the late 1980s (and in single cases even earlier) in the process engineering industry. From these days on MPC has been applied for decades predominantly in this field. Prevalent therein are rather large but slow plants to be controlled. Typically only very few instances of each plant exist. Thus, dedicated and relatively powerful computational hardware could be utilized. As a result, the achieved computation times were fast enough to enable execution at the sampling rates required by these processes. Surveys on the history of MPC applications are found for example in (Lee, 2011; Mayne et al., 2000; Qin and Badgwell, 2003).

In the year 2003, the well cited survey paper on industrial MPC technology (Qin and Badgwell, 2003) still mainly reported traditional applications of MPC in the process engineering field. Yet, it was also recognized therein that "*Significant growth areas include the chemicals, pulp and paper, food processing, aerospace and automotive industries.*" Since then, this assessment has completely turned out to be true. In (Lee, 2011) we find that "*A significant number of applications involving mechanical and electronic systems are now being reported in the literature. [...] The reported applications include vehicle traction control, suspension, direct injection stratified charge engines, ducted fan in a thrust-vectored flight control experiment, automotive powertrains, magnetically actuated mass spring damper system, power converters, multicore thermal management, and so on.*" Further surveys as well as scanning the scope of today's application oriented publications in the area of MPC

1

confirm this trend. Today, most novel applications for MPC are reported in automotive industry, see (Del Re et al., 2010; Di Cairano, 2012; Hrovat et al., 2012) and papers referenced therein. Various power electronic applications are reported and referenced in (Vazquez et al., 2014), aerospace applications are found e.g. in (Joos et al., 2012; Kang and Hedrick, 2009; Liu et al., 2011; Schlipf et al., 2013).

An increasing request for efficiency, for providing strict (safety) guarantees during operation and generally the demand to operate systems at high performance has been a driving force for the application of MPC also in these industries. MPC is perfectly suitable to satisfy these requests as has been acknowledged not only by theoretically oriented researchers but by practitioners as well. On the other hand, looking more in detail at the characteristics of such new applications, it is clear that they pose new requirements and challenges regarding the MPC algorithms to be employed. The controlled processes in novel fields are typically orders of magnitude faster than traditional ones. Consequently, the employed control algorithms have to keep up with that pace. At the same time, in contrast to classical applications of MPC, novel applications are typically produced in large volumes. This makes keeping the costs of the employed computational hardware to a minimum much more important than it used to be in classical applications and in many novel applications embedded computational hardware is to be employed. In turn, this means that fairly restricted resources in terms of computational power and memory are available for the execution of the control algorithms. This situation is alleviated to some extent by the fact that the plants in these applications can typically be modeled using few states and in many cases a linear system model is sufficient. Summarizing, in order to be able to employ MPC in novel applications and exploit its advantages in these fields, fast and efficient MPC algorithms are required which can be executed at high sampling rates even on low cost and embedded hardware.

Recent and ongoing research addresses this task from several directions and also the thesis at hand contributes towards this goal. Among the existing results, firstly tailored numerical optimization schemes are developed, see for example (Diehl et al., 2009; Domahidi et al., 2012; Wang and Boyd, 2010). Second, more attention is paid to the interplay of optimization algorithms with the (embedded) computational hardware they are run on. This is, among others, done in (Jerez et al., 2014; Zometa et al., 2012) and in the papers referenced in (Jones and Kerrigan, 2015) and in (Lucia et al., 2016). A third strategy addresses shifting some numerical effort offline in order to alleviate online computations. The most prominent representative of this class of approaches is so-called explicit MPC where, in advance, an explicit solution of the optimal control problem is pre-computed offline as a function of the state. This function is stored and made available online where it only has to be evaluated for the current system state, see for example (Alessio and Bemporad, 2009; Bemporad et al., 2002; Tøndel et al., 2003) for results regarding linear systems and (Darup and Mönnigmann, 2012; Johansen, 2004; Summers et al., 2010) for nonlinear systems. At the core of this approach is the idea to consider the finite horizon open-loop optimal control problem underlying MPC as a multi-parametric program, an optimization problem which depends on a parameter, here the system state. Despite some advantages explicit MPC definitely has, the approach is restricted to rather small MPC problems. For growing problem dimensions, the structure of the explicit solution quickly grows too complex to evaluate it efficiently online. Beyond that, even when considering small MPC problems, it is by no means clear that the most efficient online algorithm is obtained from shifting either all or none of the optimization offline. Intermediate strategies might be superior. In addition to that, in high volume applications it is especially attractive to invest a higher offline effort to simplify online computations and enable the employment of cheaper computational hardware. In this case, a higher one-time (offline) invest pays back with the savings achieved in each instance of the controlled system sold.

Thus, intermediate "semi-explicit" MPC strategies, which combine the so-far mainly separate worlds of explicit and online optimization based MPC algorithms, are an obvious and at the same time very promising extension to the available collection of MPC schemes. They have the potential to satisfy the application driven current demand for fast and efficient MPC algorithms. With respect to completely explicit approaches, clearly improved scalability in terms of the problem size can be expected; with respect to purely online optimization based schemes, a simplified numerical optimization can be expected. Thus, semi-explicit approaches can extend applicability of MPC to system classes beyond what is currently possible or beyond where the application of MPC is currently profitable. In the survey on explicit MPC (Alessio and Bemporad, 2009), this assessment is shared: *"Future research efforts should therefore pursue three main directions. [...] Third, semi-explicit methods should be also sought, in order to pre-process of line as much as possible of the MPC optimization problem without characterizing all possible optimization outcomes, but rather leaving some optimization operations on-line."* Additionally, such combination is clearly interesting from a conceptual point of view.

In this thesis, we aim at presenting a semi-explicit MPC approach which fulfills these expectations: An MPC approach is introduced which joins concepts of the explicit and of the online optimization based approach to combine their individual strengths. The resulting MPC scheme should be quickly executable and require only little computational resources. This way, the results of this thesis shall contribute to the applicability of MPC to new problem classes. At the same time, above discussed benefits of MPC should be maintained by the semi-explicit MPC scheme and system theoretic properties such as feasibility of the optimization, high control performance and asymptotic stability of the closed loop should be guaranteed.

1.2 Contributions and outline of the thesis

The basic principle of the semi-explicit approach contributed in this thesis is as follows: For a given specific MPC problem, in a first offline stage state-dependent parametrizations are computed such that they optimally approximate solutions to the MPC optimization problem. In the second online phase, these parametrizations are employed to simplify the numerical solution of the optimization. Based thereon, a simple and efficient overall online MPC scheme is formulated. A main characteristic of the proposed approach is that the parametrizations are computed data-based, namely via the application of a suitably adapted subspace clustering algorithm which is applied to optimal input sequences. Throughout the thesis this general approach will be elaborated for different problem classes so that overall the approach is applicable to linear as well as to several classes of nonlinear systems.

In the following, the contributions and the structure of this thesis is outlined more in detail.

Chapter 2 – Background on MPC and parametric optimization

In this chapter, the background required for the results of this thesis is introduced. In particular, results on setpoint stabilizing MPC and multi-parametric optimization are given.

Chapter 3 – General results on semi-explicit MPC based on subspace clustering

In this chapter, the general methods, concepts and ideas underlying the semi-explicit MPC scheme proposed in this thesis are introduced. Each in the most general setting, parametrizations are introduced, their computation is addressed and based on the parametrizations a semi-explicit MPC scheme is formulated and evaluated. Parts of this chapter are based on (Goebel and Allgöwer, 2013, 2014a,b, 2017a,b). In summary, the contributions of this chapter are as follows:

- We propose a novel type of parametrization suitable to approximate the solution of a multi-parametric optimization problem and exploit the parameter-dependence of the solution.

- We show that employing the parametrization in a parametric program, beneficial properties of the parametric program are maintained such that its numerical solution for a given value of the parameter is simplified.

- We propose a general procedure to compute the parametrization applying a tailored subspace clustering algorithm which is proposed as well.

- Based on employing the parametrization, we formulate a simple and numerically efficient semi-explicit MPC scheme for a rather general class of discrete-time systems.

- We prove closed-loop asymptotic stability, strong and recursive feasibility, and an upper bound on the closed-loop costs for the semi-explicit MPC scheme.

- We contribute two illustrative numerical examples: A first one in which the explicit solution of a multi-parametric quadratic program is recovered as a special case of the parametrization, and a second one which illustrates general aspects of the approach.

The question of how to guarantee feasibility of the parametrized optimization problem for a given set of states is left aside in this chapter, but is postponed to later chapters.

Chapter 4 – Semi-explicit MPC for linear systems

In this chapter, we particularize the general results introduced in Chapter 3 for the case of linear systems with polytopic constraints. Exploiting special properties of this problem class, several specific approaches to compute parametrizations and guarantees of further system theoretic properties are established. Parts of this chapter are based on (Goebel and Allgöwer, 2013, 2014a, 2017a,b). In particular, the contributions of this chapter are as follows:

- We propose three ways of finding parametrizations with guaranteed feasibility of the parametrized optimization problem for linear systems. Each approach has different advantages which are evaluated and compared against each other.

- We establish a theoretic upper bound of the relative cost increase incurred by applying the parametrizations.

- We introduce an offline stability check for the parametrization which allows to employ a simplified online MPC scheme.

- We present numerical examples which show applicability of the method to systems with up to an eight-dimensional state space and which illustrate tuning parameters and benefits of the method.

Chapter 5 – Almost explicit MPC for linear systems

This chapter addresses in detail the efficient numerical solution of the optimization problem obtained from applying a parametrization with scalar parameter in a linear MPC optimization problem. Parts of this chapter are based on (Goebel and Allgöwer, 2015, 2017a). More precisely, the contributions of this chapter are as follows:

- We formulate an extremely simple numerical optimization procedure which can be applied if a univariate parametrization is utilized in a linear MPC problem.

- We formulate simplifications of the efficient numerical optimization procedure made possible by identifying and removing non-relevant constraints and by exploiting properties of the parametrization.

- We assess the numerical complexity of the method and guarantee that it is simpler than evaluating a corresponding explicit solution of the same problem.

Chapter 6 – Semi-explicit MPC for nonlinear systems

In this chapter we come back to general nonlinear systems and present results and algorithms to compute offline parametrizations with guaranteed feasibility of the resulting parametrized optimization problem for classes of nonlinear problems. Parts of this chapter are based on (Goebel and Allgöwer, 2014b). In detail, the contributions of this chapter are as follows:

- We establish a test which allows to prove feasibility of a parametrized nonlinear MPC problem employing robust MPC techniques.

- We formulate a procedure to compute parametrizations with guaranteed feasibility for incrementally stable systems and systems with known Lipschitz constant of the system dynamic function.

- We propose a tailored clustering algorithm which allows to improve the shape of clusters in state space, thereby improving applicability of the general method to nonlinear problems.

- We apply the proposed method to an illustrative nonlinear numerical example for which we establish feasibility guarantees.

Chapter 7 – Conclusion

This chapter finally summarizes the results and contributions of this thesis and addresses directions for related further research.

The algorithms developed in this thesis were implemented in MATLAB (version R2016b) employing the additional tools *Multi-Parametric Toolbox 3.0* (Herceg et al., 2013), *Yalmip* (Löfberg, 2004) and *Gurobi* (Gurobi Optimization, 2016). The code was run on an Intel Core I7-3520M CPU at 2.9 GHz and using 16 GB of RAM under a Windows 10 operating system.

Chapter 2

Background on MPC and parametric optimization

In this chapter the basic problem setup as well as basic theoretical results underlying the material presented in this thesis are introduced. First, we review in the area of model predictive control some basic theoretic results. Second, we have a closer look at parametric optimization problems as they are at the core of the considered class of MPC problems and as the theory presented in this work essentially addresses efficiently solving them for fixed parameter taken from a given set of parameters.

2.1 Model Predictive Control

We consider control of nonlinear time-invariant discrete-time systems of the form

$$x_{k+1} = f(x_k, u_k), \ k \geq 0, \tag{2.1}$$

with x_0 given, where $f : \mathbb{X} \times \mathbb{U} \rightarrow \mathbb{R}^n$ holds, $x_k \in \mathbb{X} \subseteq \mathbb{R}^n$ is the system state and $u_k \in \mathbb{U} \subset \mathbb{R}^m$ is the control input at time $k \in \mathbb{N}$. The sets $\mathbb{X}$ and $\mathbb{U}$ define state and input constraints, respectively. We assume that the system has an equilibrium $\overline{x}$ in the interior of $\mathbb{X}$, that is $\overline{x} \in \text{int}(\mathbb{X})$, with corresponding input $\overline{u} \in \mathbb{U}$ such that $\overline{x} = f(\overline{x}, \overline{u})$ holds. To simplify matters we assume throughout the thesis without loss of generality that the equilibrium is at the origin, that is $\overline{x} = 0$ and $\overline{u} = 0$.

MPC for setpoint regulation

Model predictive control in a basic form addresses stabilization of a setpoint $\overline{x}$, $\overline{u}$ of a system (2.1) subject to the minimization of a given performance criterion. The performance objective is expressed via a cost function $\ell : \mathbb{X} \times \mathbb{U} \rightarrow \mathbb{R}_{\geq 0}$ which is positive definite with respect to the setpoint $\overline{x}$, $\overline{u}$ and whose sum along closed-loop trajectories is to be minimized. At the core of such basic MPC scheme is the following finite horizon open-loop optimal control problem

$$
\begin{aligned}
V^*(x_k) = \min_{U_k, X_k} \ & J_s(X_k, U_k) \\
\text{s.t. } & x_{j+1|k} = f(x_{j|k}, u_{j|k}), \ j = 0, \dots, N-1 \\
& x_{j|k} \in \mathbb{X}, \ j = 1, \dots, N-1 \\
& u_{j|k} \in \mathbb{U}, \ j = 0, \dots, N-1 \\
& x_{N|k} \in \mathbb{X}_T \\
& x_{0|k} = x_k
\end{aligned}
\tag{2.2}
$$

where

$$J_s(X_k, U_k) = \sum_{j=0}^{N-1} \ell(x_{j|k}, u_{j|k}) + V_T(x_{N|k}) \qquad (2.3)$$

holds. Therein, $U_k = [u_{0|k}^\top, \ldots, u_{N-1|k}^\top]^\top \in \mathbb{R}^{mN}$ denotes a stacked predicted input sequence at time k over a prediction horizon of $N \in \mathbb{N}$ steps. The corresponding stacked state sequence is denoted by $X_k = [x_{1|k}^\top, \ldots, x_{N|k}^\top]^\top \in \mathbb{R}^{nN}$. Furthermore, $\mathbb{X}_T \subseteq \mathbb{X}$ and $V_T : \mathbb{X}_T \to \mathbb{R}_{\geq 0}$ are a terminal constraint set and a terminal cost function, respectively.

In order to ensure existence of a minimum of Optimization (2.2), further assumptions are required. Different sets of assumptions are sufficient to guarantee this existence, of which we use the following rather intuitive ones as standing assumptions throughout this thesis.

Assumption 2.1. *Let the functions f, ℓ and V_T be continuous. Let the sets $\mathbb{X}$ and $\mathbb{X}_T$ be closed and let the set $\mathbb{U}$ be compact.*

Without further assumptions, Optimization (2.2) clearly might have a non-unique solution. Whenever this is relevant, we assume that a unique choice among the solutions is made.[1] By $U_k^*(x_k) = [u_{0|k}^{*}{}^\top(x_k), \ldots, u_{N-1|k}^{*}{}^\top(x_k)]^\top$ we denote the minimizing input sequence of Optimization (2.2). The standard MPC algorithm based on this optimization is then as follows.

Algorithm 1 Basic MPC algorithm

1: obtain current state x_k
2: solve Optimization (2.2)
3: apply $u = u_{0|k}^*(x_k)$ to the plant and return to step 1

Application of this algorithm results in the closed-loop system

$$x_{k+1} = f(x_k, u_{0|k}^*(x_k)). \qquad (2.4)$$

Considering that the goal is stabilization of the origin, relevant questions concerning this scheme are: i) If the optimization is feasible at time k for the state x_k, is it also feasible at time $k+1$ for the closed-loop state x_{k+1}? This property is called *recursive feasibility*. ii) Is the origin an asymptotically stable equilibrium of the closed-loop system (2.4)? iii) If so, what is the region of attraction of the origin? In order to be able to give satisfactory answers to these questions, some further assumptions regarding the setup are required.

Assumption 2.2. *Let ℓ be positive definite in both of its arguments.*

Assumption 2.3 (Mayne et al. (2000)). *Assume that the terminal constraint set $\mathbb{X}_T$, the terminal control law κ and the terminal cost function V_T are such that the following holds:*

1. *The terminal constraint set $\mathbb{X}_T$ fulfills $\mathbb{X}_T \subseteq \mathbb{X}$, it is closed and $0 \in \mathbb{X}_T$.*

2. *There is a terminal control law $\kappa : \mathbb{X}_T \to \mathbb{R}^m$, $x \mapsto \kappa(x)$ such that $\kappa(x) \in \mathbb{U}$ for all $x \in \mathbb{X}_T$.*

---

[1]For the results in this thesis, we implicitly rely on some regular structure of the optimizer of (2.2) as a function of the state x. If the minimizer is unique, such regular structure is given. In non-unique cases, we assume that the selection function used to obtain a unique solution preserves this structure.

3. *The terminal constraint set is invariant under the terminal control law, i.e., for all* $x \in \mathbb{X}_T$ *it holds that* $f(x, \kappa(x)) \in \mathbb{X}_T$.

4. *The terminal cost function* V_T *satisfies* $V_T(x) - V_T(f(x, \kappa(x))) \geq \ell(x, \kappa(x))$ *for all* $x \in \mathbb{X}_T$ *and* $V_T(0) = 0$.

The following result holds.

Theorem 2.1 (Mayne et al. (2000)). *Let Assumptions 2.2 and 2.3 hold. Algorithm 1 is recursively feasible, the origin is an asymptotically stable equilibrium of the closed-loop system* (2.4), *the region of attraction is the set of states for which Optimization* (2.2) *is feasible.*

The main idea for the proof of the latter theorem is as follows. Due to the assumptions on the stage cost function ℓ, the optimal value function V^* fulfills the properties of a Lyapunov candidate function. If the optimization has been solved before, in each time step a candidate solution U_C to the optimization exists consisting of the shifted previous predicted input sequence with the input obtained via the terminal control law appended. The cost of this candidate solution is an upper bound on the cost of the optimal solution. Assumptions 2.2 and 2.3 ensure that a certain cost decrease of the candidate solution is achieved, which implies that the optimal value function V^* fulfills the properties of a Lyapunov function. In fact results are available which make use of a weaker version of Assumption 2.2. For example, positive definiteness of ℓ with respect to an output function together with a corresponding detectability condition are sufficient to prove asymptotic stability of the loop closed with the above scheme. More information on the computation of a suitable terminal cost and terminal constraint set can be found in (Chen and Allgöwer, 1998) for the continuous time case and in (Rawlings and Mayne, 2009) for the discrete-time version.

Under some slightly stronger assumptions on the MPC ingredients, the requirements for Algorithm 1 can be relaxed such that not necessarily an optimal solution is required therein but the results remain valid if initially any feasible input sequence is chosen which is then improved along closed-loop trajectories, c.f. (Scokaert et al., 1999).

Further approaches to guarantee closed-loop stability

Besides the discussed approach to achieve and prove stability of the closed-loop system, a variety of different concepts exists for this purpose, see the overview in (Mayne et al., 2000). Dual mode schemes basically use an MPC algorithm as introduced above to drive the system state into a region around the origin and therein switch to a local control law exploiting its stabilizing properties. Variable horizon schemes reduce the prediction horizon length during operation, thereby automatically ensuring a decrease of the value function. Contractive and stability enforced MPC schemes ensure closed-loop stability employing in some way an additional control Lyapunov function which has to be known explicitly a priori. Unconstrained MPC schemes do not employ any terminal constraints and, nevertheless, allow to formulate closed-loop stability guarantees for sufficiently long prediction horizons, typically under additional assumptions on the system to be controlled, see (Grüne, 2012) for an overview.

Basically all of these schemes have at their core an optimization problem which depends on and online has to be solved for the current system state. These problems are structurally similar to Optimization (2.2) but with additional and/or omitted terms in the objective function and/or constraints.

MPC addressing tasks beyond setpoint stabilization

Indeed there is a plethora of results on MPC schemes addressing a variety of problem classes other than setpoint stabilization. Many of these schemes fit into a setup similar or even identical to the one considered so far. Using an extended system, the following cases can be treated in a setup exactly as considered above: Tracking of a piecewise constant reference signal, incorporating integral action in the control law, imposing rate constraints on the control signal and attenuating known disturbances acting on the system. In (Bemporad et al., 2002) these cases are considered for linear systems whereas (Mayne, 2014) addresses the general nonlinear case. Enforcing mixed state and input constraints of the form $(x, u) \in \mathbb{Z} \subset \mathbb{X} \times \mathbb{U}$ and relaxing the constraints to soft constraints can be achieved via rather straight forward adjustments of Optimization (2.2).

Robust and stochastic MPC schemes address control of an uncertain or disturbed system when either bounds or stochastic information is available about the disturbance and/or the uncertainty, see for example (Kouvaritakis and Cannon, 2015; Mayne, 2015) and references therein. Economic MPC schemes directly address a closed-loop economic objective expressed via the stage cost function used in the optimization. The main difference from the setpoint stabilizing schemes is that the stage cost function in economic MPC is not necessarily positive definite with respect to any admissible setpoint of the controlled system, see (Ellis et al., 2014) for an overview. Further existing MPC schemes consider distributed problems and algorithms, control of switched and hybrid systems, periodic systems, time-delay systems, linear parameter-varying systems and many more. Also many of these schemes possess at their core an optimization similar to Optimization (2.2).

In this thesis, we will focus on setpoint stabilizing schemes as a prototypical application of the theory developed. In many cases extension of the results to schemes beyond that is straight forward or possible via slight adaptations.

Linear MPC

Let us come back to the basic setpoint stabilizing MPC scheme introduced above and have a closer look at particularities of the most important subclass of systems contained therein, namely linear systems. MPC for this problem class is addressed in depth e.g. in (Borrelli et al., 2015). Here, the general nonlinear System (2.1) has the form of a linear system

$$x_{k+1} = Ax_k + Bu_k, \ k \geq 0, \tag{2.5}$$

with x_0 given and matrices $A \in \mathbb{R}^{n \times n}$ and $B \in \mathbb{R}^{n \times m}$. State and input constraint sets are typically assumed to be polytopes, i.e., they can be represented as

$$\mathbb{X} = \{x | F_x x \leq f_x\}, \tag{2.6}$$

$$\mathbb{U} = \{u | F_u u \leq f_u\} \tag{2.7}$$

with suitable $F_x \in \mathbb{R}^{n_x \times n}$, $f_x \in \mathbb{R}^{n_x}$, $F_u \in \mathbb{R}^{n_u \times m}$ and $f_u \in \mathbb{R}^{n_u}$ where n_x and n_u is the number of state and input constraints, respectively. Furthermore, typically a quadratic stage cost

$$\ell(x, u) = x^\top Q x + u^\top R u \tag{2.8}$$

with symmetric, positive definite matrices $Q \in \mathbb{R}^{n \times n}$ and $R \in \mathbb{R}^{m \times m}$ is used. The corresponding unconstrained infinite horizon optimal control input is given as linear state feedback $u = K_{\mathrm{LQR}}x$, where $K_{\mathrm{LQR}} \in \mathbb{R}^{m \times n}$ can be determined via the solution $P_{\mathrm{ARE}} \in \mathbb{R}^{n \times n}$ of the discrete-time Algebraic Riccati Equation corresponding to (2.5) and (2.8). Using the terminal control law $\kappa(x) = K_{\mathrm{LQR}}x$ and the terminal cost

$$V_T(x) = x^\top P_{\mathrm{ARE}}x, \tag{2.9}$$

a corresponding terminal constraint set can be determined as the so-called *maximal output admissible set* with respect to the 'output' $y = K_{\mathrm{LQR}}x$ (Gilbert and Tan, 1991). This set is the maximal invariant set of the system $x^+ = (A + BK_{\mathrm{LQR}})x$ for which state constraints are satisfied and the terminal control law fulfills the input constraints. Thus, for the given problem terminal ingredients can readily be computed which are optimal in the sense that for states in the terminal set the terminal control law is optimal and the terminal cost equals the infinite horizon cost-to-go and the terminal set is the largest such set which is invariant under the terminal control law. Furthermore, under the given assumptions, the terminal constraint set is also a polytope

$$\mathbb{X}_T = \{x | F_T x \leq f_T\} \tag{2.10}$$

with $F_T \in \mathbb{R}^{n_T \times n}$ and $f_T \in \mathbb{R}^{n_T}$ with n_T the number of terminal constraints. We will refer to such terminal cost, terminal control law and terminal constraints as terminal ingredients according to the unconstrained linear quadratic regulator (LQR) setup. An important benefit of restricting attention to this class of problems is the resulting simplicity of Optimization (2.2) which will become more obvious after a reformulation of the problem which is to be introduced below.

2.2 Parametric optimization

2.2.1 The parametric optimization problem in MPC

Let us next have a closer look at Optimization (2.2) at the core of the basic MPC scheme in Algorithm 1. Noticing that this problem is static and time-independent, its essence can be captured in a much simpler formulation. In fact the constraints and the objective function of the optimization depend, aside from the decision variables, on the system state x for which the optimization is to be solved. Such optimization problem which depends on a parameter which lives in a multi-dimensional parameter space, here the state x, is called a *multi-parametric program* (mpP).

In order to obtain a compact formulation of the optimization, predicted state variables can be eliminated from the decision variables expressing them as a function of the initial condition and the system inputs. This step called *condensing* yields an equivalent optimization in the sense that both problems possess the same predicted input sequences as optimizers from which the corresponding state sequence can be recovered uniquely.[2]

Using these insights, we can formulate a general multi-parametric program which captures the structure and basic properties of the original optimization problem. We assume that the

---

[2]This equivalence does not hold with respect to applicability and properties of numerical solution strategies, see for example (Axehill, 2015; Diehl et al., 2009).

constraints of (2.2) can be formulated in terms of level sets of explicitly given continuous functions, that is in the form $\tilde{g}(X, U) \leq 0$ where $\tilde{g} : \; \mathbb{X}^N \times \mathbb{U}^N \to \mathbb{R}^r$ is known and continuous, $r \in \mathbb{N}$ is the number of constraints and the inequality is to be read element-wise. Constraints for the condensed problem can then be expressed as $g(x, U) \leq 0$ where $g : \mathbb{X} \times \mathbb{U}^N \to \mathbb{R}^r$ and g is obtained from $\tilde{g}$ eliminating therein predicted states via f. The objective function is reformulated as $J : \; \mathbb{X} \times \mathbb{U}^N \to \mathbb{R}$, yielding the rather general mpP

$$V^*(x) = \min_U J(x, U)$$
$$\text{s.t. } g(x, U) \leq 0. \tag{2.11}$$

Therein, in addition the time index k is omitted and the notation $U = [u_0^\top, \dots, u_{N-1}^\top]^\top$ is used for the predicted input sequence. The functions g and J are compositions of continuous functions and, thus, are also continuous. We use the following notation for admissible solutions and for the set of feasible states.

Definition 2.2. *Let* $\mathbb{U}_N(x) = \{U \in \mathbb{R}^{mN} | g(x, U) \leq 0\}$ *be the set of all N-step input sequences which are compatible with the constraints of mpP (2.11) for state x.*

Definition 2.3. *Let* $\mathbb{X}_N = \{x \in \mathbb{R}^n | \mathbb{U}_N(x) \neq \varnothing\}$ *be the set of states for which a solution to mpP (2.11) exists. We call $\mathbb{X}_N$ the N-step controllable set to the terminal set.*

From a control perspective, $\mathbb{U}_N(x)$ is the set of all N-step input sequences, which drive state x into the terminal set respecting state and input constraints and $\mathbb{X}_N$ is the set of states which can be driven into the terminal set within N steps respecting state and input constraints.

We now return to the linear case as introduced above. Consider a linear system (2.5) with polytopic state, input and terminal constraints (2.6), (2.7) and (2.10). Then the general mpP (2.11) has the form

$$\min_U J(x, U)$$
$$\text{s.t. } GU \leq Ex + W \tag{2.12}$$

for suitable matrices $G \in \mathbb{R}^{r \times mN}$, $E \in \mathbb{R}^{r \times n}$ and $W \in \mathbb{R}^r$. It results that the constraints are jointly convex in x and in U and that $\mathbb{U}_N(x)$ and $\mathbb{X}_N$ are polytopic sets. If in addition one assumes a quadratic stage cost (2.8) and a terminal cost (2.9), mpP (2.11) has the form

$$\min_U U^\top H U + x^\top F U + x^\top Y x$$
$$\text{s.t. } GU \leq Ex + W \tag{2.13}$$

for $H = H^\top \in \mathbb{R}^{mN \times mN}$, positive definite, $F \in \mathbb{R}^{n \times mN}$ and $Y \in \mathbb{R}^{n \times n}$. These results are readily obtained by algebraic reformulations.

Definition 2.4 (mpQP). *A multi-parametric program of the form (2.13) is called multi-parametric quadratic program (mpQP).*

Furthermore, if the stage and terminal cost in the MPC problem formulation have the form $\ell(x, u) = \|Qx\|_p + \|Pu\|_p$ and $V_T(x) = \|Px\|_p$ for $p \in \{1, \infty\}$, the mpP can be reformulated as

$$\min_{\hat{U}} c^\top \hat{U}$$
$$\text{s.t. } G\hat{U} \leq Ex + W \tag{2.14}$$

for $c \in \mathbb{R}^{\tilde{n}}$, $G \in \mathbb{R}^{r \times \tilde{n}}$, $E \in \mathbb{R}^{r \times n}$ and $W \in \mathbb{R}^r$, with $\tilde{n} \geq mN$.

Definition 2.5 (mpLP). *A multi-parametric program of the form* (2.14) *is called multi-parametric linear program (mpLP).*

For the case $p = 1$ again simple algebraic calculations yield the result and $\tilde{n} = mN$ holds whereas in the case $p = \infty$ additional decision variables have to be introduced for the reformulation, yielding $\tilde{n} = (m+1)N + N + 1$.

As has become apparent in the foregoing discussions, a parameter-dependent optimization problem is at the core of most MPC formulations. The mpP reformulation (2.11) of Optimization (2.2) emphasizes this parameter-dependence. For an MPC scheme, the resources and the time required to obtain a solution of this problem online for a given state x are the crucial characteristics which decide about applicability of the MPC scheme. In traditional MPC schemes, the parameter-dependence was rarely exploited or accounted for,[3] but typically the problem was simply formulated for the current system state rendering it a non-parameter-dependent nonlinear program which was then solved numerically. Then many results followed the obvious idea to compute or approximate the state-to-input map defined by the optimization explicitly offline and evaluate it online to control the system. More recently, also results became available which follow intermediate strategies by exploiting parameter-dependence of the optimization problem to simplify its numerical solution online. We will discuss these results in more detail below comparing them to the results presented in this thesis.

2.2.2 Applications of parametric optimization

Even though model predictive control is probably the largest field of application for parametric programming, mpPs appear and are applicable beyond MPC. Parametric programming approaches can be used to solve bilevel (and more generally multilevel) optimization problems. In bilevel optimization problems, the objective and/or constraints of an upper level problem depend on the solution of a second lower-level optimization problem which, in turn, is affected by the solution of the upper level problem. Solving the lower level problem as mpP reduces the upper level problem to a standard optimization problem. Examples can be found in (Faísca et al., 2007; Ryu et al., 2004; Zhou and Spanos, 2016) and below in this thesis (in Subsection 4.4.3). Another type of applications is in multi-objective optimization problems, i.e., optimization problems which posses multiple conflicting objectives. Here the parameters are used to weight these objectives against each other. A parameter-dependent solution in this case characterizes the Pareto frontier and enables to easily evaluate the problem for different weightings of the objectives. Applications of multi-objective optimization are found among others in the medical and in the financial sector, see e.g. the examples presented in (Romanko, 2010). Combinations of the applications mentioned so far and also applications beyond that exist. Parametric programming can be applied also in (global) numerical optimization algorithms and some directly application motivated optimization problems are best formulated in a parameter-dependent fashion.

---

[3]Warm-starting the optimization using a (shifted) previous solution is a common strategy which might be considered a simple basic way of exploiting parameter-dependence.

2.2.3 Properties of parametric optimization problems

Parametric programs have been studied theoretically and some theoretical foundation independent of the applications mentioned is available. See for example (Bank et al., 1983; Fiacco, 1984; Guddat et al., 1990; Hogan, 1973) and, for a summary of the most important results in an MPC context, (Borrelli et al., 2015). Results on the properties of the (generally point to set) mapping $x \mapsto \mathbb{U}_N(x)$, of the value function mapping $x \mapsto V(x)$ and of the (possibly set-valued) optimizer mapping $x \mapsto U^*(x)$ under different assumptions on the properties of the parametric programs have been investigated.

The most relevant question for our application is whether or not these mappings behave well enough to allow to exploit the parameter-dependence and to approximate the corresponding functions sufficiently well with reasonable effort. A priori this is not clear and in fact some surprisingly irregular cases can be constructed. Without going into tedious technical details, we cite the bottom line of a relevant beneficial result (Diehl et al., 2009): "*It is a well-known fact from parametric optimization, cf. (Guddat et al., 1990), that the solution manifold has smooth parts when the active set does not change (and bifurcations are excluded), but that non-differentiable points occur whenever the active set changes.*" *Active set* denotes the set of indices of active constraints at an optimal solution, i.e., rows of the constraint function g of (2.11) which are fulfilled with equality. The *solution manifold* consists of points satisfying the Karush-Kuhn-Tucker (KKT) conditions corresponding to the mpP (2.11) and contains all local optimizers. To turn this into a precise statement, further technicalities would have to be regarded. The conclusion, nevertheless, remains valid that under reasonable assumptions there is a mapping $x \mapsto U^*(x)$ which behaves (at least piecewise) well and can be approximated by techniques to be developed in this thesis.

Quadratic and linear program with affine constraints

Much more can be said about the solution structure of multi-parametric linear programs and multi-parametric quadratic programs as they arise from linear MPC problems. To characterize properties of these parametric programs, we need the following definitions.

Definition 2.6 (Piecewise affine & piecewise quadratic function)**.** *Let* $\mathbb{P} \subset \mathbb{R}^n$ *and* $\mathcal{P}_1, \ldots, \mathcal{P}_{n_{CR}}$ *be a polyhedral partition of* $\mathcal{P}$.

- *A function* $h : \mathbb{P} \to \mathbb{R}^m$ *is called piecewise affine (PWA) if it can be written as*

$$h(x) = C_j x + D_j \text{ for } x \in \mathcal{P}_j \tag{2.15}$$

  where $C_j \in \mathbb{R}^{m \times n}$, $D_j \in \mathbb{R}^n$, $j = 1, \ldots, n_{CR}$.

- *A function* $h : \mathbb{P} \to \mathbb{R}$ *is called piecewise quadratic (PWQ) if it can be written as*

$$h(x) = x^\top F_j x + G_j^\top x + K_j \text{ for } x \in \mathcal{P}_j \tag{2.16}$$

  where $F_j \in \mathbb{R}^{n \times n}$, $G_j \in \mathbb{R}^n$ *and* $K_j \in \mathbb{R}$, $j = 1, \ldots, n_{CR}$.

The following results about mpLPs and mpQPs are well known and can be found for example in (Bank et al., 1983; Borrelli et al., 2015).

Theorem 2.7 (Properties of mpLP). *Consider a multi-parametric linear program* (2.14). *The feasible set* $\mathbb{X}_N$ *is a polytopic set. The value function* $x \mapsto V^*(x)$ *is continuous, convex and piecewise affine over* $\mathbb{X}_N$. *If the optimizer* $U^*(x)$ *is unique for all* $x \in \mathbb{X}_N$, *then the optimizer function* $x \mapsto U^*(x)$ *is continuous and piecewise affine. Otherwise it is always possible to define a continuous and piecewise affine optimizer function.*

Theorem 2.8 (Properties of mpQP). *Consider a multi-parametric quadratic program* (2.13) *and let* $\left(\begin{smallmatrix} 2H & F^\top \\ F & 2Y \end{smallmatrix} \right) \succeq 0$. *Then the feasible set* $\mathbb{X}_N$ *is a polytopic set, the value function* $x \mapsto V^*(x)$ *is continuous, convex and piecewise quadratic over* $\mathbb{X}_N$, *the optimizer* $U^*(x)$ *is unique and the optimizer function* $x \mapsto U^*(x)$ *is continuous and piecewise affine.*

In these cases the function $x \mapsto U^*(x)$ is explicitly computable and we refer to this function as *the explicit solution* of an mpLP or mpQP, respectively. The condition $\left(\begin{smallmatrix} 2H & F^\top \\ F & 2Y \end{smallmatrix} \right) \succeq 0$ is equivalent to joint convexity of the mpQP in x and U and is typically fulfilled by mpQPs resulting from MPC problems (Bemporad et al., 2002). The polytopes on which the optimizer function is affine are called critical regions and they are defined via related active sets of the corresponding mpLP or mpQP, respectively. The results formulated in the latter theorems are the foundation of explicit MPC, see (Bemporad et al., 2002; Tøndel et al., 2003) and many subsequent publications.

Chapter 3

General results on semi-explicit MPC based on subspace clustering

In the general introduction above, we discussed that there is a large and still increasing demand for fast and computationally efficient MPC schemes. In this chapter we are going to introduce our general method to formulate such fast and efficient MPC algorithms. The general method is at the core of this thesis and will also form the starting point for subsequent chapters.

This chapter is partly based on (Goebel and Allgöwer, 2013, 2014a,b, 2017a,b).

3.1 Introduction and problem formulation

In this chapter we address stabilization of the origin of a nonlinear discrete-time system

$$x_{k+1} = f(x_k, u_k), \ \ k \geq 0 \tag{3.1}$$

with given initial condition x_0, and which is subject to state and input constraints

$$x_k \in \mathbb{X} \subseteq \mathbb{R}^n, \ \ u_k \in \mathbb{U} \subset \mathbb{R}^m$$

and a given performance criterion in form of a stage cost function ℓ. A setpoint stabilizing MPC scheme is to be used for this purpose which employs a terminal cost and a terminal constraint, that is, we consider an MPC scheme based on the finite horizon open-loop optimal control problem (2.2). Regarding the constraint sets and the stage cost function, we keep Assumptions 2.1 and 2.2 on closure, compactness and continuity. Regarding terminal constraint, terminal cost and corresponding terminal control law in the MPC setup, we keep Assumption 2.3 and require in addition that the terminal control law does not only exist but is explicitly known.

Assumption 3.1. *Let Assumption 2.3 hold and assume in addition that the terminal control law κ therein is explicitly known.*

In the following, we will work based on the reformulation of the MPC optimization problem as multi-parametric program

$$V^*(x) = \min_U J(x, U)$$
$$\text{s.t. } g(x, U) \leq 0. \tag{3.2}$$

Our goal in this chapter is to propose a general method to simplify online numerical solution of mpP (3.2) for a fixed value of x and employ this in an MPC scheme such that the overall scheme is simplified compared to the basic MPC scheme formulated in Algorithm 1 and system theoretic guarantees can still be established. This will be achieved by employing a parametrization which maps a lower-dimensional parameter $\tilde{U}$ to the original decision variable U and in addition accounts for and exploits state-dependence of optimal solutions of mpP (3.2). Online optimization is then carried out over the lower-dimensional decision variable $\tilde{U}$.

Within this chapter, first the parametrizations will be introduced and their application to mpPs will be discussed. Second, a method to compute the parametrization via a suitable data mining algorithm such that it optimally approximates the solutions to a given mpP is introduced. Third, a so-called semi-explicit MPC scheme based on employing the parametrization is formulated and system theoretic guarantees for the scheme are established. Finally, we evaluate the general method in examples and theoretically and relate it to existing simplified MPC schemes.

As this chapter mainly serves the purpose to introduce the general method and its structural properties, answers to some questions, which require to be addressed more problem specific, will be postponed to later chapters of this thesis. Explicitly not covered within this chapter is the issue of evaluating or achieving feasibility of the parametrized optimization problem for a given set of states. Here, instead, it will be assumed that a certain set of states is known for which the parametrized optimization is feasible. The question of how to adjust a parametrization such that feasibility of the parametrized optimization for a given set of states is guaranteed will be a central aspect in Chapter 4 for linear systems and in Chapter 6 for classes of nonlinear systems.

In order to avoid confusion of the parameters of the mpP with the parameters introduced by the parametrization, we will refer to the mpP parameters x as *states* and keep the term *parameter* reserved for those parameters introduced by the parametrization. Nevertheless, this should not hide the facts that even in an MPC application quantities beyond states can play the role of parameters in the mpP and that the basic approach is applicable to general mpPs beyond those arising in MPC.

3.2 The parametrizations and their application to multi-parametric programs

Next, the suggested type of parametrization will be introduced and its application to mpPs will be examined.

3.2.1 The parametrizations

The parametrizations we propose map a parameter onto an optimizer of the considered mpP and, beyond that, depend also on the state to be applied for. Thus, they are a mapping from the state space and the parameter space to the space of optimizers of the mpP. Yet, we restrict our attention to parametrizations of a certain structure. On the one hand, this structure ensures that beneficial properties are carried over from the original optimization problem to the parametrized optimization problem. On the other hand, parametrizations which possess this structure can be computed in a particularly efficient way.

Definition 3.1. *We define the general parametrization considered here as a collection of* K *mappings*

$$
p = \Big\{ p_1 : \ \hat{\mathcal{D}}_1 \times \mathbb{R}^q \to \mathbb{R}^{mN}, \quad (x, \tilde{U}) \mapsto p_1(x, \tilde{U}) = M_1 \tilde{U} + N_1 \phi(x),
$$
$$
\ldots \tag{3.3}
$$
$$
p_K : \ \hat{\mathcal{D}}_K \times \mathbb{R}^q \to \mathbb{R}^{mN}, \quad (x, \tilde{U}) \mapsto p_K(x, \tilde{U}) = M_K \tilde{U} + N_K \phi(x) \Big\}
$$

where $\{ M_i \in \mathbb{R}^{mN \times q}, \ N_i \in \mathbb{R}^{mN \times b}, \ i = 1, \ldots, K \}$ *are matrices,* $\phi : \mathbb{R}^n \to \mathbb{R}^b$ *is a function and* $\{ \hat{\mathcal{D}}_i \subset \mathbb{R}^n, \ i = 1, \ldots, K \}$ *are (generally overlapping) sets which fulfill for a given set* $\mathcal{X}_f \subset \mathbb{R}^n$ *in state space*

$$
\mathcal{X}_f \subseteq \bigcup_{i=1}^{K} \hat{\mathcal{D}}_i. \tag{3.4}
$$

Choosing a partition $\{\mathcal{D}_1, \ldots, \mathcal{D}_K\}$ *of* $\mathcal{X}_f$ *which is compatible with the parametrization in the sense that* $\mathcal{D}_i \subseteq \hat{\mathcal{D}}_i$ *holds for* $i = 1, \ldots, K$, *the parametrization can be turned into a well-defined mapping* $p : \mathcal{X}_f \times \mathbb{R}^q \to \mathbb{R}^{mN}$, $(x, \tilde{U}) \mapsto p(x, \tilde{U})$ *given by*

$$
p(x, \tilde{U}) = \begin{cases} p_1(x, \tilde{U}) & \text{if } x \in \mathcal{D}_1 \\ \vdots & \vdots \\ p_K(x, \tilde{U}) & \text{if } x \in \mathcal{D}_K. \end{cases} \tag{3.5}
$$

In most of the situations encountered below, it turns out convenient to use overlapping sets $\hat{\mathcal{D}}_i$ and leave a choice of the sets $\mathcal{D}_i$ open in advance. Keeping in mind that any (explicit or implicit) choice of a partition turns (3.3) immediately into (3.5), we will in the sequel refer to both (3.3) and (3.5) as parametrization.

The fact that this structure of the parametrization is rather simple and yet flexible has a number of advantages which will be evaluated below. Additionally, the structure of the parametrization covers relevant special cases: The choice $\phi(x) = [x^\top \ 1]^\top$ results in parametrizations of the form $p_i(x, \tilde{U}) = M_i \tilde{U} + \mathcal{K}_i x + a_i$ with $\mathcal{K}_i \in \mathbb{R}^{mN \times n}$ and $a_i \in \mathbb{R}^{mN}$ which will play an important role in the application to linear problems. Beyond that, the limiting case of completely omitting the M_i matrices in the parametrizations is conceptually interesting. Whenever we set $q = 0$ below, we refer to this case with slight abuse of notation. Choosing in this case K sufficiently large, this covers the explicit solution of mpLPs and mpQPs making the parametrizations structurally relevant.

The considered parametrization p is defined by two types of quantities. The first type of quantities adjusts structural properties. It comprises the number of parts K, the dimension q of the space the parameter $\tilde{U}$ lives in and the function ϕ including the dimension b of its co-domain. We call these quantities *hyperparameters*, adopting this term from the machine learning community, see e.g. (Duan et al., 2003). The second type of quantities are the *values* for the matrices $\{ M_i$ and $N_i, \ i = 1, \ldots, K \}$ and the sets $\{ \mathcal{D}_i, \ i = 1, \ldots, K \}$ which tailor the parametrization for a particular problem. Whereas the hyperparameters are tuning knobs to be chosen and adjusted by a user, the remaining quantities are determined in an automated procedure for given hyperparameters and a given problem. Both will be addressed below in depth.

3.2.2 Applying the parametrizations in multi-parametric programs

It is a well known and widely exploited fact that the numerical solution of an optimization problem can be simplified by reducing the dimension of its decision variable while the number of constraints is left unchanged or is decreased and beneficial structural properties of the optimization problem are maintained. In many MPC applications this is achieved by applying move-blocking techniques which parametrize the predicted input sequence in a way such that some consecutive inputs are "blocked" to be constant. Here the goal is to build on the same principle but to go far beyond the existing results by applying much more powerful parametrizations. Additional simplifications become possible by exploiting state-dependence of optimal solutions. The underlying idea here remains the same: Replacing the original decision variable U in the mpP (3.2) by the parametrization p so that optimization can be carried out over the lower-dimensional parameter $\tilde{U}$ as new decision variable. We define the parametrized mpP as follows.

Definition 3.2. *For $x \in \hat{\mathcal{D}}_i$ the i-th parametrized optimization and its value function are defined as*

$$\boldsymbol{P}_{p_i}(x): \qquad V_{pi}^*(x) = \min_{\tilde{U}} J(x, p_i(x, \tilde{U}))$$
$$\text{s.t. } g(x, p_i(x, \tilde{U})) \leq 0.$$

For a well-defined overall parametrization (3.5), the overall parametrized optimization and its value function are defined via

$$\boldsymbol{P}_p(x) = \boldsymbol{P}_{p_i}(x) \text{ and } V_p^*(x) = V_{pi}^*(x) \text{ for } x \in \mathcal{D}_i.$$

A few important questions arise immediately when parametrizing the decision variable:

- How does the parametrization affect feasibility of the parametrized optimization?

- How does the parametrization affect the achievable optimal value?

- How does the parametrization affect structural properties of the optimization which, in turn, influence how efficiently the optimization can be solved numerically?

These questions will be addressed in the following in a general non-problem specific way and we will reconsider them more in detail specific for the respective problem classes in subsequent chapters.

As the parametrization maps (for fixed state x) a lower-dimensional parameter $\tilde{U}$ into a higher-dimensional co-domain, it is immediately clear that (for fixed state x) the image of the parametrization is a strict subset of the co-domain. Application of the parametrization in the mpP can be seen as a restriction of the search space for solutions to this image. Due to the presence of constraints in the mpP, this restriction might lead to infeasibility of the parametrized optimization, i.e., there will generally exist states x for which the original mpP is feasible but the parametrized optimization is infeasible. Furthermore, the optimal value of the parametrized optimization will generally be larger than that of the original problem, i.e., $V^*(x) \leq V_p^*(x)$. Whereas it is clearly desirable to achieve low objective function values even using a parametrization, guaranteeing feasibility of the parametrized optimization for a certain set of states is absolutely mandatory for the method to be applicable. Throughout this thesis, we will generally take the perspective that a set $\mathcal{X}_f$ is given for which the parametrized optimization $\mathbf{P}_p(x)$ has to be feasible and a corresponding suitable parametrization is to be found. We call $\mathcal{X}_f$ the *desired feasible set* and define a feasible parametrization as follows.

Definition 3.3. *Consider a multi-parametric program* (3.2) *and a parametrization p of type* (3.5) *with resulting parametrized multi-parametric programs* $\{\mathbf{P}_{p_i}, \ i = 1, \ldots, K\}$. *Parametrization p is called feasible on the set* $\mathcal{X}_f \subseteq \mathbb{X}_N$ *or in short a* feasible parametrization *if there exist sets* $\{\hat{\mathcal{D}}_1, \ldots, \hat{\mathcal{D}}_K \subset \mathbb{R}^n\}$, *such that*

$$\mathcal{X}_f \subseteq \bigcup_{i=1}^{K} \hat{\mathcal{D}}_i \tag{3.6}$$

and

$$\mathbf{P}_{p_i}(x) \ \text{is feasible for all } x \in \hat{\mathcal{D}}_i. \tag{3.7}$$

Clearly, any choice for the sets $\mathcal{D}_i$ used in (3.5) which fulfills $\mathcal{D}_i \subseteq \hat{\mathcal{D}}_i$ will ensure that $\mathbf{P}_p(x)$ is feasible for all states x in $\mathcal{X}_f$.

In order to evaluate structural properties of the parametrized optimization, let us have a closer look at the image of the parametrization. For a fixed value of the state x, $p(x, \cdot)$ is an affine mapping and its image is a q-dimensional affine subspace in $\mathbb{R}^{mN}$. Loosely speaking, the mappings obtained by composing the objective function $J(x, \cdot)$ and the constraint function $g(x, \cdot)$ with this affine mapping, respectively, inherit beneficial properties of J and g, respectively. In particular, continuity (and possibly continuous differentiability) of the constraint and the value function in U are transferred to continuity of the corresponding parametrized function with respect to $\tilde{U}$. Thus, Assumption 2.1 is still sufficient to guarantee existence of a minimum of the parametrized optimization. Beyond that, employing this affine map ensures that convexity properties of the original mpP (3.2) are inherited by the parametrized mpP $\mathbf{P}_p(x)$.

Proposition 3.4. *Consider an mpP* (3.2) *and its parametrized version* $\mathbf{P}_{p_i}$.

- *If the objective function of mpP* (3.2) *is convex for fixed value of x, so is the objective function of the parametrized mpP* $\mathbf{P}_{p_i}(x)$.

- *If the set of admissible solutions* $\mathbb{U}(x)$ *of mpP* (3.2) *is convex for a fixed value of x, so is the set of admissible solutions* $\{\tilde{U}|g(x, p_i(x, \tilde{U})) \leq 0\}$ *of the parametrized mpP* $\mathbf{P}_{p_i}(x)$.

- *The number of non-redundant constraints of the parametrized mpP* $\mathbf{P}_{p_i}$ *is less or equal to the number of non-redundant constraints of the mpP* (3.2).

Proof. The first two statements both follow from the fact that a convex function composed with an affine function is again a convex function (Boyd and Vandenberghe, 2009). The third statement is immediately clear from the definition of the constraints. $\qquad\square$

As the dimension of the decision variable is reduced, the number of constraints is not increased and typically also reduced and advantageous properties of a given optimization problem are maintained, numerically solving the parametrized optimization problem will generally be simpler than solving the original mpP.

As discussed, employing a parametrization inherently deteriorates feasibility and performance of the parametrized optimization. On the other hand, a suitable tailored parametrization mitigates these effects. It ensures that the parametrized mpP is feasible for a large set of states and it results in only slightly increased optimal values for feasible states. As the solutions of the mpP are state-dependent, it is reasonable to employ a parametrization which accounts for this dependence by also depending itself on the state to be applied for. The parametrization proposed here possesses this property and depends on the state in two ways: First, the parametrization is defined piecewise such that each piece is best suitable for a specific region in state space. Second, within each piece the term $N_i\phi(x)$ incorporates a state-dependent offset.

Summarizing, the structure of the proposed parametrization is simple enough to maintain some important beneficial properties in the parametrized mpP and it is general enough to account for the dependence of optimal solutions to the mpP on the parameter x in two ways. Furthermore, due to its structure, it can be computed efficiently as will become obvious in the next section.

3.3 Finding the parametrizations

As discussed above, the goal is to have a parametrization which results in a feasible parametrized mpP for a large set of states and which achieves a low objective function value for the feasible states. A main challenge when computing a parametrization is that a priori neither the sets $\hat{\mathcal{D}}_i$ are known, nor the corresponding values of M_i and N_i. As these parts closely depend on each other, a suitable approach is required to determine these quantities in a joint procedure.

The core idea behind our approach to find such parametrization is to work data-based. In particular a set of states $\{x_1, \ldots, x_s\}$ is sampled from the feasible set of the original mpP and corresponding optimal solutions $\{U^*(x_r), r = 1, \ldots, s\}$ of the mpP are computed. By employing a tailored data mining algorithm, the parametrization is then determined such that it optimally approximates these solutions. Relying on the fact that the parameter-to-optimizer mapping of the mpP is sufficiently well-behaved, this parametrization is also suitable for states close to the states explicitly used.

When computing a parametrization, first the hyperparameters, i.e., the values K, q, and the function ϕ have to be chosen. These quantities mainly decide about the complexity of the parametrization and about its flexibility to approximate solutions of the original mpP. In a general machine learning application, choosing hyperparameters is a delicate task and extensive research within the machine learning community has addressed careful selection thereof. The goal there is to have a model which is flexible enough to fit the structural properties of the considered data but which is restrictive enough such that it does not over-fit the training data used, see for example (Duan et al., 2003). In our application, this aspect can be an issue but will generally not be the deciding aspect. For our purpose, the hyperparameters are mainly used to adjust the complexity of the parametrization as this translates into complexity and "the degree of explicitness" of the resulting semi-explicit

MPC scheme. Generally, the goal is to choose the complexity of the parametrization low which directly translates into low values of K and q and a simple function ϕ. Selection of the hyperparameters will be addressed below in more detail. For the moment, assume that a set of hyperparameters has been chosen. The remaining degrees of freedom M_i, N_i and $\hat{\mathcal{D}}_i$, $i = 1, \ldots, K$ will then be determined data-based employing a tailored subspace clustering algorithm. We next briefly introduce the general concept of subspace clustering before we turn to the employed tailored algorithm more in detail.

3.3.1 Subspace clustering

Clustering algorithms generally address the task of grouping a given set of data points into several clusters in a way such that similarity among the data points within each cluster is high according to a given similarity measure. A typical similarity measure is to consider the distance of the data points to a cluster specific centroid. In the simplest case the centroid is a point in the space of data points. This clustering problem is addressed by the well-known K-means clustering algorithm as proposed in (MacQueen et al., 1967). In subspace clustering, loosely speaking the similarity measure is the distance of the data points to a cluster specific subspace, i.e., the centroid here is a cluster specific subspace. To be more precise, subspace clustering addresses the following type of problems: Given a set of data points $\{U_r \in \mathbb{R}^{mN}, r = 1, \ldots, s\}$, group the data into a predefined number K of clusters such that the data in each cluster lies close to a linear or affine subspace of given dimension q. The subspaces are also to be found. This means the goal within each cluster is to find a matrix $M_i \in \mathbb{R}^{mN \times q}$ and possibly a constant $a_i \in \mathbb{R}^{mN}$ such that for all U_r assigned to the cluster $U_r \approx M_i \tilde{U}_r + a_i$ holds for a suitable $\tilde{U}_r \in \mathbb{R}^q$.

Different types of subspace clustering algorithms exist, see for example (Vidal, 2011) for an overview. An important class among them are iterative algorithms. These algorithms can be seen as derivatives of the K-means clustering algorithm. The general principle behind these algorithms is to solve the original complex clustering problem via iteratively solving two simple problems: Starting from an initial assignment, iteratively first update the cluster centroid based on the data assigned to the cluster and then update cluster membership by assigning each data point to the closest centroid. One example of an iterative subspace clustering algorithm is the so-called K-q-flats clustering method reported in (Bradley and Mangasarian, 2000).

3.3.2 A tailored subspace clustering algorithm

We first illustrate why the problem of finding a parametrization of the proposed type based on data is closely related to subspace clustering problems and then introduce our customized subspace clustering algorithm. Assume for the moment that the parametrization does not contain any state-dependent information, i.e., assume $\phi(x) \equiv 0$ or $\phi(x) \equiv$ const.. The image of the parametrization then has the form $\{\text{span}(M_i) + a_i, i = 1, \ldots, K\}$, i.e., it consists of K linear ($a_i = 0$) or, respectively, affine ($a_i \neq 0$) q-dimensional subspaces. Our goal is to choose the subspaces such that they approximate the data given, i.e., the optimizers $\{U^*(x_r), r = 1, \ldots, s\}$, optimally on average. Obviously, this is exactly the type of problem which subspace clustering addresses. Re-incorporating state-dependence of the parametrization via general functions $\phi(x)$ adds a regression aspect beyond the pure subspace clustering. Recall that in subspace clustering centroids are sought which are linear or affine subspaces of the space the data lies in. In our case, the centroids consist of affine subspaces and additionally linear mappings in between both parts of a data point.

Objective of the algorithm

For the situation considered here, we assume that a source of information for each data point U_r is available in form of a corresponding variable $y_r \in \mathbb{R}^b$. Thus, the proposed method deals with pairs of data points $\{(y_r, U_r) \in \mathbb{R}^b \times \mathbb{R}^{mN}, \ r = 1, \ldots, s\}$ and approximates only the second components U_r. In particular, beyond what is done in pure subspace clustering, here within each cluster a linear correlation of all y_r and corresponding U_r in the cluster is to be established. Loosely speaking, within each cluster the goal is to find matrices M_i and N_i such that for all U_r assigned to the cluster $U_r \approx M_i \tilde{U}_r + N_i y_r$ holds for a suitable $\tilde{U}_r \in \mathbb{R}^q$. For the remainder of this section, the reader may think of $y_r = x_r$ as a simple choice (Goebel and Allgöwer, 2014a) and of $y_r = \phi(x_r)$ as the most general version we are aiming for. Hence, for each cluster a combined low rank approximation (corresponding to the subspaces)/least square fitting problem (corresponding to the linear correlation) is solved. In more detail, for a given set of data points $\{(y_r, U_r) \in \mathbb{R}^b \times \mathbb{R}^{mN}, \ r = 1, \ldots, s\}$ and parameters $K \in \mathbb{N}$ and $q \in \mathbb{N}$, the method aims at finding a solution to the optimization problem

$$\min_{M_i, N_i, \tilde{U}_r, \mu_i^r, \ i=1,\ldots,K, \ r=1,\ldots,s} J_C$$
$$\text{s.t. } \mu_i^r \in \{0, 1\} \text{ and } \sum_{i=1}^{K} \mu_i^r = 1, \tag{3.8}$$

where

$$J_C = \sum_{i=1}^{K} \sum_{r=1}^{s} \mu_i^r \|U_r - M_i \tilde{U}_r - N_i y_r\|_2 \tag{3.9}$$

with $M_i \in \mathbb{R}^{mN \times q}$, $N_i \in \mathbb{R}^{mN \times b}$ and $\tilde{U}_r \in \mathbb{R}^q$. Therein, μ_i^r determines cluster membership of the data point pairs, i.e., data point pair r is assigned to cluster i if $\mu_i^r = 1$. The matrices N_i establish a linear correlation of y-parts and U-parts, the matrices M_i define the q-dimensional subspace the residuals $U_r - N_i y_r$ lie close to. The variables $\tilde{U}_r$ are only needed for a compact problem formulation. Note that not the matrices M_i are defined uniquely in the given optimization but generally only their images are.

Procedure

Due to its complexity, optimization problem (3.8) does not lend itself to a direct solution procedure. Better suitable is an iterative strategy as mentioned above. The two sub-problems employed therein are as follows:

Cluster update step: Within the cluster update step, optimal cluster specific matrices M_i, N_i are to be found for a given assignment of points to the clusters. This is achieved via for each cluster i solving the optimization problem

$$\min_{M_i, N_i, \tilde{U}_r} \sum_{\{r | \mu_i^r = 1\}} \|U_r - M_i \tilde{U}_r - N_i y_r\|_2. \tag{3.10}$$

As stated in (Gabriel, 1978), this combined least squares/low rank approximation problem can be solved via two separate steps. First, the least squares computation yields $N_i = \hat{U}_i(\hat{Y}_i)^+$, where $\hat{U}_i$ and $\hat{Y}_i$ are matrices collecting the U_r variables and the y_r variables, respectively, currently assigned to cluster i and $(\cdot)^+$ denotes the Moore-Penrose pseudo-inverse. Second, the matrix M_i is taken as the first q principal components of the residual $\hat{U}_i - N_i\hat{Y}_i$. The $\tilde{U}_r$ variables are not computed explicitly.

Cluster assignment step: During the assignment step, the parameters μ_i^r are to be optimized based on fixed cluster specific matrices M_i, N_i. Hence, μ_i^r is taken as the argument of a solution of the optimization problem

$$\min_{\mu_i^r, \tilde{U}_r} \sum_{i=1}^{K} \sum_{r=1}^{s} \mu_i^r \|U_r - M_i\tilde{U}_r - N_iy_r\|_2$$
$$\text{s.t. } \mu_i^r \in \{0,1\} \text{ and } \sum_{i=1}^{K} \mu_i^r = 1. \tag{3.11}$$

Note that this simply means to assign each one of the data point pairs to the cluster by which it is best approximated. Typically we consider rather low values of K, so this optimization can be solved efficiently via an exhaustive enumeration strategy.

Using these two steps, we can formulate the tailored subspace clustering algorithm in Algorithm 2. The following result regarding convergence of the algorithm holds.

Algorithm 2 Tailored subspace clustering algorithm

Input: pairs $\{(y_r, U_r) \in \mathbb{R}^b \times \mathbb{R}^{mN}, \ r = 1, \ldots, s\}$, parameters $K, q \in \mathbb{N}$
1: initialize μ_i^r, $i = 1, \ldots, K$, $r = 1, \ldots, s$ randomly subject to the constraints in (3.8)
2: update matrices M_i, N_i via (3.10), for $i = 1, \ldots, K$
3: update μ_i^r, $i = 1, \ldots, K$, $r = 1, \ldots, s$ via (3.11)
4: **if** μ_i^r has changed w.r.t. previous iteration **then**
5:     go back to step 2
6: **else**
7:     done
8: **end if**
Output: matrices M_i, N_i, $i = 1, \ldots, K$, cluster assignment μ_i^r, $i = 1, \ldots, K$, $r = 1, \ldots, s$

Proposition 3.5. *If Algorithm 2 terminates, it terminates at a locally optimal solution.*

Here "locally optimal" solution means that for this solution the objective function J_C can neither be decreased by changing the assignment of points to clusters alone nor by changing the parameters of the clusters alone.

Proof. Assume the algorithm has terminated at a certain assignment, which due to termination has appeared repeatedly in the last two iterations. Changing the assignment alone would not decrease J_C as the last step of the algorithm was to choose an optimal assignment. Starting from a given assignment, the cost J_C after execution of step 2 is unique. So the cost after step 2 was executed in between the two repeated occurrences of the final assignment equals the cost achievable if after the final assignment step 2 is executed again. As the cost can never increase while running the algorithm, also the cost after termination equals this cost and the final solution is locally optimal. ☐

In practice, the given formulation of the algorithm together with the latter proposition has turned out useful as in literally all considered numerical tests after few iterations the same assignment appeared repeatedly, terminating the algorithm. Nevertheless, it cannot be guaranteed that at a locally optimal solution an assignment appears repeatedly in two *consecutive* iterations. As is typical for iterative clustering algorithms, convergence to a locally optimal solution after a finite number of iterations (independently of termination of the algorithm) can be shown, see for example (Bradley and Mangasarian, 2000). The idea of the convergence proof is to show that J_C can never increase while running the algorithm and considering that there is only a finite number of different assignments, J_C has to reach after a finite number of iterations a value which is not decreased any further. Theoretically, there might exist loops of length greater than one on constant level sets of J_C where each element of the loop is locally optimal. In such case the proposed stopping criterion would be insufficient. In order to theoretically guarantee termination at a locally optimal solution it would be sufficient to terminate the algorithm whenever any assignment appears repeatedly at all (not necessarily in two consecutive time steps). As this would require to store in the worst case K^s different assignments (K the number of clusters and s the number of data points used), this is impractical. Sacrificing convergence to a locally optimal solution, a maximum number of iterations could be used as an additional verifiable stopping criterion to guarantee termination.

3.3.3 Computing the parametrizations data-based

To compute the parametrizations data-based, first representative states $\{x_r, \ r = 1, \ldots, s\}$ are sampled from the feasible set of the mpP to be parametrized. For each training state x_r the mpP is solved to obtain the optimizer $U_r = U^*(x_r)$. The states are then mapped to $y_r = \phi(x_r)$ and the subspace clustering algorithm is applied to the pairs y_r, U_r. This yields the matrices M_i and N_i and cluster membership of the data point pairs. Cluster membership of the data is then used to define the sets $\hat{\mathcal{D}}_i$ and $\mathcal{D}_i$ in state space. In order to guarantee feasibility of the parametrized optimization, typically a postprocessing step refining the parametrization will be required. This is disregarded here for the moment and we only address computation of *raw* parametrizations, which are guaranteed to be feasible only in special situations. Algorithm 3 formalizes the procedure. Refinement of the parametrizations to guarantee feasibility will be a main subject of Chapters 4 and 6 of this thesis.

Algorithm 3 Computation of raw parametrization

Input: mpP (3.2), parameters $K, q \in \mathbb{N}$, function ϕ, training states $\{x_1, \ldots, x_s\}$
1: solve mpP (3.2) to obtain $U_r = U^*(x_r)$, for $r = 1, \ldots, s$
2: compute $y_r = \phi(x_r)$, for $r = 1, \ldots, s$
3: run Algorithm 2 with (y_r, U_r), $r = 1, \ldots, s$, K and q as input Let M_i, N_i, $i = 1, \ldots, K$ and μ_i^r, $r = 1, \ldots, s$, $i = 1, \ldots, K$, be the output of Algorithm 2
Output: M_i, N_i, $i = 1, \ldots, K$, and μ_i^r, $r = 1, \ldots, s$, $i = 1, \ldots, K$

Finding the sets $\mathcal{D}_i$: Depending on the problem details and on the choice of training states, different strategies to define the sets $\{\mathcal{D}_i, \ i = 1, \ldots, K\}$, based on the output of Algorithm 3 are best suitable. If the training data has been chosen from a regular grid, assigning each point $x \in \mathcal{X}_f$ to the same cluster as the closest training point is reasonable. This yields the sets

$$\mathcal{D}_i = \{x \in \mathcal{X}_f | \text{closest training state to } x \text{ is assigned to cluster } i\} \qquad (3.12)$$

for $i = 1, \ldots, K$. In cases where there is more than one closest training state any selection function is needed to achieve uniqueness. As the training states have been sampled from a regular grid, determining membership of a point $x \in \mathcal{X}_f$ to one of the sets $\mathcal{D}_i$ is numerically simple even without having any explicit representation of the sets. If the training states have not been sampled from a regular grid, finding the closest one for a given point $x \in \mathcal{X}_f$ will generally be computationally more expensive. In this case, application of a multi-class support vector machine (mSVM) (Hsu and Lin, 2002) to define and also identify cluster membership will generally be simpler. Training states labeled with their cluster membership could be used to train the mSVM and the mSVM would define the sets $\mathcal{D}_i$ implicitly via its output. Further options are possible and will be discussed at a later stage of this thesis.

Selection of the training data: Loosely speaking, in the proposed approach the training states used should in some sense cover and represent the part of the state space for which a parametrization is sought. A reasonable approach is to sample the states from a regular grid covering the relevant area in state space. Some perturbation can be added to such states to avoid having any artificial structure in the data. If the shape of the desired feasible set in state space is degenerated, using a rotated grid can yield better results. In higher-dimensional state spaces a more careful selection of the states might become necessary to keep the number of states manageable.

Choosing initial conditions for the clustering: Iterative clustering algorithms of the type employed here typically only converge to local minima. In order to mitigate this effect, re-running the algorithm from different initial conditions and choosing the initial conditions in an elaborate manner are common strategies, see for example (Bradley and Fayyad, 1998). Here, reasonable initial conditions can be chosen based on the application considered and based on the fact that typically states which are close to each other yield similar solutions of the mpP. First, randomly select K out of the training states of which each one will be used to initialize on of the clusters. To each selected training state add sufficiently many nearby training states so that M_i and N_i for the corresponding cluster can be initialized uniquely. Based thereon, an initial cluster assignment is found.

3.4 The semi-explicit MPC algorithm

In this section, the online part of the semi-explicit MPC algorithm will be introduced and its system theoretic properties will be examined.

3.4.1 The basic semi-explicit online MPC algorithm

We start stating a basic semi-explicit MPC scheme which employs a parametrization of the proposed type to simplify online computations. Originating from this scheme, variants thereof will be derived below in this section and in later chapters of this thesis. The goal here is to formulate an algorithm which exploits the simplified parametrized online optimization and, at the same time, is equipped with system theoretic guarantees. For the latter purpose, the existence of an admissible candidate solution to the optimization problem which is obtained by shifting the previous predicted input sequence by one time step and appending the input obtained applying the terminal control law to the final predicted state is crucial.

For the algorithm, we assume that a parametrization p of type (3.5) is known which ensures that $\mathbf{P}_p(x)$ is feasible for all $x \in \mathcal{X}_f$ where $\mathcal{X}_f \subset \mathbb{R}^n$. The basic semi-explicit MPC scheme is formulated in Algorithm 4. In the semi-explicit MPC scheme in each

---

Algorithm 4 Semi-explicit online MPC scheme

Input: System (3.1), $\mathbf{P}_{p_i}$, $i = 1, \ldots, K$, parametrization p, terminal control law κ, $\tau \in (0, 1]$
1: initialize $J_{\text{comp}} = \infty$ and $U_C = 0 \in \mathbb{R}^{mN}$
2: **loop**
3:      obtain current state x
4:      **if** $x \in \mathcal{X}_f$ **then**
5:          find i^* such that $x \in \mathcal{D}_{i^*}$
6:          solve $\mathbf{P}_{p_{i^*}}(x)$, get $\tilde{U}^*$
7:          set $U = p_{i^*}(x, \tilde{U}^*)$
8:          **if** $J_{\text{comp}} < J(x, U)$ **then**
9:             set $U = U_C$
10:          **end if**
11:      **else**
12:          set $U = U_C$
13:      **end if**
14:      set $u_{\text{se}}(x) = [I_m \ 0]U$ and apply $u_{\text{se}}(x)$ to the plant
15:      set $J_{\text{comp}} = J(x, U) - \tau \ell(x, u_{\text{se}}(x))$
16:      set $U_C = [u_1^\top, \ldots, u_{N-1}^\top, \kappa(x_N)^\top]^\top$
17: **end loop**

---

time-step first feasibility of the parametrized optimization is tested via evaluating if the current state is contained in the feasible set $\mathcal{X}_f$. If this is the case a simplified parametrized optimization problem is solved for the current state and if the obtained solution yields a sufficiently low predicted open-loop cost, the newly optimized input is applied. Otherwise, the candidate solution is used. In Algorithm 4 the tuning parameter τ was introduced to adjust whether to prefer newly optimized solutions or to prefer solutions with low predicted costs. Choosing τ large generally reduces closed-loop costs whereas choosing τ small weights the feedback aspect stronger by preferring new solutions. In the nominal case where no disturbances act on the system and the system to be controlled behaves exactly according to the model used, the candidate sequence will always be feasible and ensure a sufficient cost decrease. However, in the non-nominal case, i.e. plant-model mismatch and/or disturbances are present, incorporating feedback via re-computing input sequences is important. Straight-forward extensions which include the candidate solution U_C into the optimization will be discussed below.

3.4.2 System theoretic properties of the basic semi-explicit MPC scheme

Next, system theoretic properties of the basic semi-explicit MPC scheme are examined.

Asymptotic stability

In order to establish asymptotic stability for a loop closed via Algorithm 4 we assume continuity of the value function $V_p^*(x)$ at the origin.

Assumption 3.2. *Let the value function $V_p^*(x)$ of the parametrized multi-parametric program $\boldsymbol{P}_p$ be continuous at the origin.*

We will establish a result which implies satisfaction of this assumption below. First, the main result regarding asymptotic stability is stated.

Theorem 3.6 (Closed-loop stability). *Consider System (3.1) with corresponding MPC formulation resulting in mpP (3.2). Let p be a parametrization such that the corresponding parametrized mpP $\boldsymbol{P}_p(x)$ is feasible for all $x \in \mathcal{X}_f$. Let Assumptions 2.1, 2.2 and 3.1 regarding the MPC setup hold, let Assumption 3.2 regarding the parametrized mpP hold. Then, Algorithm 4 is recursively feasible, it renders the origin of the closed-loop system asymptotically stable, and the set $\mathcal{X}_f$ is a subset of the region of attraction of the closed loop.*

Proof. First note that for $x \in \mathcal{X}_f$ the algorithm is initially feasible as $\boldsymbol{P}_p(x)$ is feasible. Recursive feasibility follows from the existence of admissible candidate input sequences. To show asymptotic stability define $J_{x_0}^{\text{ol}}(x_k) = J(x_k, U)$ where x_k is the k-th state of the closed-loop state trajectory originating from x_0 and applying Algorithm 4 and U is the chosen predicted input sequence at time step k defined in Algorithm 4. Along closed-loop trajectories, the relaxed dynamic programming inequality

$$J_{x_0}^{\text{ol}}(x_{k+1}) \leq J_{x_0}^{\text{ol}}(x_k) - \tau \ell(x_k, u_k) \quad \text{for } k \geq 0 \tag{3.13}$$

holds. This follows from the fact that due to Assumption 2.3 candidate input sequences ensure fulfillment of this inequality (with $\tau = 1$) and newly optimized sequences are applied only if they fulfill this inequality.

Using this relation, convergence can be shown along the lines of (Scokaert et al., 1999): Due to the properties of the stage cost function, $J_{x_0}^{\text{ol}} \geq 0$ holds and due to (3.13), $J_{x_0}^{\text{ol}}(x_k)$ decreases along closed-loop trajectories. It follows that $J_{x_0}^{\text{ol}}(x_{k+1}) - J_{x_0}^{\text{ol}}(x_k) \to 0$ for $k \to \infty$, which implies that $\ell(x_k, u_k) \to 0$ and thus $x_k \to 0$ for $k \to \infty$.

Next, we prove stability. Let $x_0 \in \mathcal{X}_f$, then we have

$$V^*(x_k) \leq J_{x_0}^{\text{ol}}(x_k) \leq V_p^*(x_0) \quad \text{for } k \geq 0. \tag{3.14}$$

Assumption 2.3 implies that $V^*(x)$ is continuous at the origin and $V_p^*(x)$ is continuous at the origin by assumption. Hence, both functions are upper bounded by $\mathcal{K}$-functions in an open neighborhood of the origin. Thus we have i) for all $\epsilon > 0$ there exists an $\bar{\epsilon} > 0$ such that $\mathcal{A}_{\bar{\epsilon}} = \{x \in \mathbb{R}^n | V^*(x) \leq \bar{\epsilon}\} \subseteq \mathcal{B}_\epsilon(0)$ and ii) for all $\bar{\epsilon} > 0$ there exists an $\underline{\epsilon} > 0$ with $\underline{\epsilon} \leq \bar{\epsilon}$ such that $\mathcal{A}_{\underline{\epsilon}}^p = \{x \in \mathbb{R}^n | V_p^*(x) \leq \underline{\epsilon}\} \subseteq \mathcal{A}_{\bar{\epsilon}}$ and iii) there exists a $\delta > 0$ such that $\mathcal{B}_\delta(0) \subseteq \mathcal{A}_{\underline{\epsilon}}$. Thus, due to (3.14) for $x_0 \in \mathcal{B}_\delta(0)$ we have that $\underline{\epsilon} \geq V_p^*(x_0) \geq J_{x_0}^{\text{ol}}(x_k) \geq V^*(x_k)$ along closed-loop trajectories. This means for $x_0 \in \mathcal{B}_\delta(0)$ it holds that $x_k \in \mathcal{A}_{\bar{\epsilon}} \subseteq \mathcal{B}_\epsilon(0)$ for all $k \geq 0$. $\qquad\square$

Note that closed-loop trajectories originating from $x_0 \in \mathcal{B}_\delta(0)$ might leave $\mathcal{A}_{\underline{\epsilon}}^p$ as $J_{x_0}^{\text{ol}}(x_k) < V_p^*(x_k)$ is possible but will remain inside $\mathcal{A}_{\bar{\epsilon}}$ as $J_{x_0}^{\text{ol}}(x_k) \not< V^*(x_k)$ holds.

Remark 3.7. *The situation considered here is slightly different from the situation considered in the well known* feasibility implies stability *result (Scokaert et al., 1999). Whereas in (Scokaert et al., 1999) the basic assumption is that initially any feasible input sequence is used and the predicted open-loop cost along closed-loop trajectories is decreased, here the predicted open-loop costs can in addition be upper bounded in each time step depending on the state. This is due to the assumption that whenever one of the optimization problems is feasible for the current state, at least one of the feasible optimization problems is solved to optimality.*

For the latter theorem, Assumption 3.2 on continuity of the value function $V_p^*(x)$ at the origin was required.

Proposition 3.8. *Consider a parametrization p of type (3.5) and let Assumptions 2.1, 2.2 and 2.3 hold. Let there be a set $\mathcal{D}_{i^0}$ such that the origin is in the interior of $\mathcal{D}_{i^0}$ and let $\tilde{U}^* = 0$ be the unique minimizer of $J(0, p_{i^0}(0, \tilde{U}))$ and let the set of minimizers of $J(x, p_{i^0}(x, \tilde{U}))$ be uniformly compact[1] near $x = 0$. Let $\phi(x)$ be continuous at the origin and let $p(0, 0) = 0$. Then, $V_p^*(x)$ is continuous at the origin.*

Proof. There is an $\epsilon_1 > 0$ such that $\mathcal{B}_{\epsilon_1}(x = 0) \subset \mathcal{D}_{i^0}$ and p is continuous on $\mathcal{B}_{\epsilon_1}(0) \times \mathbb{R}^q$. As $(x = 0, u = 0)$ is an equilibrium of the system and zero is in the interior of the state and input constraint sets, a solution of the mpP for $x = 0$ is given by $U^*(x = 0) = 0$ and at this solution no constraints are active, i.e. $g(0, 0) < 0$. This solution is optimal as $\ell \geq 0$, $\ell(0, 0) = 0$ and V_T attains its minimum at $x = 0$. Due to continuity of g, there is an $\epsilon_2 > 0$ such that for all $(x, U) \in \mathcal{B}_{\epsilon_2}(x = 0, U = 0)$ all constraints are inactive, $g(x, U) < 0$. Due to local continuity of p, there is an ϵ_3 with $\epsilon_1 > \epsilon_3 > 0$, such that for all $(x, \tilde{U}) \in \mathcal{B}_{\epsilon_3}(x = 0, \tilde{U} = 0)$ all constraints are inactive, $g(x, p(x, \tilde{U})) < 0$. The optimizer of the unconstrained version of the parametrized mpP $\overline{U}^*(x) = \arg\min_{\tilde{U}} J(x, p(x, \tilde{U}))$ is a continuous function of the parameter at $x = 0$ (due to uniqueness of the minimizer at $x = 0$ and uniform compactness of the set of minimizers near $\overline{x} = 0$) (Hogan, 1973). Thus, there is an $\epsilon_4 > 0$ such that for $x \in \mathcal{B}_{\epsilon_4}(x = 0)$ the parameter-optimizer pair $(x, \overline{U}^*(x))$ is contained in $\mathcal{B}_{\epsilon_3}(x = 0, \tilde{U} = 0)$. For (at least) all $x \in \mathcal{B}_{\epsilon_4}(x = 0)$ the constrained and the unconstrained parametrized mpP are equivalent and possess a continuous value function $V_p^*(x)$. □

Recursive feasibility

According to Theorem 3.6, the basic semi-explicit MPC algorithm remains feasible if it is initialized at a state for which the parametrized optimization is feasible. Yet, this does not generally imply that also the parametrized optimization remains feasible as the state trajectory might leave the feasible set of the parametrized optimization $\mathcal{X}_f$. Nevertheless, stronger types of recursive feasibility can be established for the algorithm under suitable conditions.

Definition 3.9 (Strong feasibility, (Gondhalekar and Imura, 2010; Kerrigan, 2001)). *An MPC problem is called strongly feasible if and only if from every feasible state the closed-loop state trajectory due to* any *sequence of feasible solutions to the MPC problem remains within the feasible set.*

---

[1] A point-to-set map $\mathcal{U}(x)$ is said to be uniformly compact near $\overline{x}$ if there exists a neighborhood $\mathcal{B}$ of $\overline{x}$ such that the closure of the set $\cup_{x \in \mathcal{B}} \mathcal{U}(x)$ is compact.

Clearly, strong feasibility implies recursive feasibility of the optimization in the sense that if the MPC algorithm is initialized such that the optimization is feasible and in the MPC algorithm optimal solutions are applied, the optimization remains feasible in all time steps.

Lemma 3.10. *Let $\mathbb{X}_{N-1}$ be the $N-1$-step controllable set to the terminal set. If $\mathbb{X}_{N-1} \subseteq \mathcal{X}_f$ then $\mathcal{X}_f$ is invariant using any feasible solution of P_p in Algorithm 4. In particular, the MPC scheme is strongly feasible.*

Proof. Let $x \in \mathcal{X}_f$, then applying the first step of any admissible input sequence the next state x^+ lies inside $\mathbb{X}_{N-1}$ as there exists a candidate sequence which drives x^+ into the terminal set within $N-1$ steps. Thus, $x^+ \in \mathbb{X}_{N-1} \subseteq \mathcal{X}_f$. This argument applies recursively. □

Below in this thesis, we will address design of parametrizations which are feasible on a given set $\mathcal{X}_f$. This can be employed to ensure $\mathbb{X}_{N-1} \subseteq \mathcal{X}_f$. Furthermore, recursive feasibility of the parametrized optimization holds if the set of initial conditions is restricted suitably.

Proposition 3.11. *Let $c > 0$ be such that $\{x|V^*(x) \leq c\} \subseteq \mathcal{X}_f$. If Algorithm 4 is initialized with $x \in \{x|V_p^*(x) \leq c\}$ the parametrized optimization in Algorithm 4 is feasible in each time step.*

Proof. Due to (3.14), the closed-loop trajectory will remain within $\{x|V^*(x) \leq c\}$ which implies that the state will remain inside $\mathcal{X}_f$ and the parametrized optimization remains feasible. □

Both sets in the latter statement are in general hard to find. In some special cases this can still be possible. Examples are linear systems with simple cost functions and structurally sufficiently simple parametrizations. Beyond that, over and under approximations of the sets might be computable.

An upper bound on the closed-loop costs

Applying an MPC scheme, one is generally interested in achieving high control performance expressed in terms of the stage cost function. For given closed-loop state and input trajectories $x_0 = x, x_1, \ldots$ and $u_0, u_1, \ldots$, closed-loop control performance is then expressed via the summed stage costs along the closed-loop trajectories,

$$J^{cl}(x) = \sum_{k=0}^{\infty} \ell(x_k, u_k). \tag{3.15}$$

As is well known, applying a nominal MPC scheme with the same type of ingredients as considered here, this closed-loop cost can be directly upper bounded by the open-loop costs

$$J^{cl}_{\text{nom}}(x) \leq V^*(x). \tag{3.16}$$

In the case of applying the basic semi-explicit MPC scheme of Algorithm 4, the closed-loop cost is generally larger and the available upper bound depends on the parameter τ used in the algorithm.

Proposition 3.12. *An upper bound on the closed-loop cost applying the basic semi-explicit MPC scheme of Algorithm 4 is given by*

$$J_{seMPC}^{cl}(x) \leq \frac{1}{\tau} V_p^*(x).$$

The proof idea is adopted from a similar result found e.g. in (Grüne and Pannek, 2011).

Proof. From the relaxed dynamic programming inequality (3.13) one obtains

$$J_{x_0}^{ol}(x_k) - J_{x_0}^{ol}(x_{k+1}) \geq \tau \ell(x_k, u_k). \tag{3.17}$$

Summing over $k \geq 0$, identifying $J_{x_0}^{ol}(x_0) = V_p^*(x_0)$ and dividing by τ yields the claim. $\square$

Even though the result gives only an upper bound on the closed-loop cost, it illustrates the effect of the tuning parameter τ. Decreasing the value of τ lets the upper bound on the closed-loop cost become arbitrarily large. A low value of τ might mean to switch to a new solution along closed-loop trajectories, even though the corresponding predicted open-loop cost is rather large and the switching increases the closed-loop costs. On the other hand, having a parametrization which admits high open-loop control performance along closed-loop trajectories renders the parameter τ ineffective and yields high control performance despite the value of τ and despite the switching.

3.4.3 Varieties of the semi-explicit algorithm

Two straight forward extensions of the basic semi-explicit MPC scheme are considered next.

Semi-explicit MPC including the candidate solution in the optimization

In the semi-explicit MPC scheme proposed so far, candidate input sequences were not explicitly included into the optimization but were used only as a fall-back solution for cases where the parametrized optimization had become infeasible or yielded insufficient solutions. In many existing parametrization based MPC schemes, the parametrization is adjusted in each time step to explicitly include the candidate sequence in its image, see for example (Hara and Kojima, 2012; Li et al., 2013; Mendez et al., 2000; Shekhar and Manzie, 2015; Valencia-Palomo and Rossiter, 2011). The candidate solution immediately yields recursive feasibility and a sufficient cost decrease along closed-loop trajectories to ensure closed-loop asymptotic stability. The downside of this approach is that the dimension of the decision variable is increased and that the optimization problem to be solved in each time-step does not only depend on the current state as a parameter but also changes depending on the current candidate input sequence. This can complicate matters and can render some simplifications to be introduced below impossible.

If the employed parametrization is a linear mapping of a parameter to the predicted input sequences, the candidate sequence can directly be used as additional basis vector for the image of the parametrization. For the parametrizations used here, in addition the non-parameter-dependent offset has to be compensated for. Two obvious choices are as follows.

Definition 3.13. *Let a parametrization p of type (3.5) and a vector $U_C \in \mathbb{R}^{mN}$ be given. We define an* extended parametrization of the first type *as $\tilde{p}_{U_C} : \mathbb{X}_N \times \mathbb{R}^{q+1} \to \mathbb{R}^{mN}$, $(x, \tilde{U}) \mapsto \tilde{p}_{U_C}(x, \tilde{U})$, with*

$$\tilde{p}_{U_C}(x, \tilde{U}) = \begin{cases} \tilde{p}_{U_C,1}(x, \tilde{U}) = [M_1 \ (U_C - N_1\phi(x))]\tilde{U} + N_1\phi(x) & \text{if } x \in \tilde{\mathcal{D}}_1 \\ \vdots \qquad\qquad \vdots & \vdots \\ \tilde{p}_{U_C,K}(x, \tilde{U}) = [M_K \ (U_C - N_K\phi(x))]\tilde{U} + N_K\phi(x) & \text{if } x \in \tilde{\mathcal{D}}_K. \end{cases} \quad (3.18)$$

Therein, $\{M_i, \ N_i, \ i = 1, \dots, K\}$, and ϕ are taken from p, and $\{\tilde{\mathcal{D}}_1, \dots, \tilde{\mathcal{D}}_K\}$ is a segmentation of $\mathbb{X}_N$ such that $\mathcal{D}_i \subseteq \tilde{\mathcal{D}}_i$ for $i = 1, \dots, K$, holds.
We define an extended parametrization of the second type *as $\tilde{\tilde{p}}_{U_C} : \mathbb{X}_N \times \mathbb{R}^{q+2} \to \mathbb{R}^{mN}$, $(x, \tilde{U}) \mapsto \tilde{\tilde{p}}_{U_C}(x, \tilde{U})$, with*

$$\tilde{\tilde{p}}_{U_C}(x, \tilde{U}) = \begin{cases} \tilde{\tilde{p}}_{U_C,1}(x, \tilde{U}) = [M_1 \ N_1\phi(x) \ U_C]\tilde{U} & \text{if } x \in \tilde{\mathcal{D}}_1 \\ \vdots \qquad\qquad \vdots & \vdots \\ \tilde{\tilde{p}}_{U_C,K}(x, \tilde{U}) = [M_K \ N_K\phi(x) \ U_C]\tilde{U} & \text{if } x \in \tilde{\mathcal{D}}_K, \end{cases} \quad (3.19)$$

where for $M_i, \ N_i, \ \tilde{\mathcal{D}}_i$ for $i = 1, \dots, K$, and ϕ the same as above holds.

Clearly, it holds that $\operatorname{im}(p(x,\cdot)) \subset \operatorname{im}(\tilde{p}(x,\cdot))$ and $\operatorname{im}(p(x,\cdot)) \subset \operatorname{im}(\tilde{\tilde{p}}(x,\cdot))$. Thus, the feasible set of the extended parametrizations is generally enlarged over the feasible set of the original parametrizations. This makes it reasonable to define the extended parametrization also beyond the domain of the original parametrization p. Based on these parametrizations, a slightly simplified online semi-explicit MPC scheme is formulated in Algorithm 5. Therein,

---

Algorithm 5 Semi-explicit MPC scheme using extended parametrization

---

Input: System (3.1), mpP (3.2), parametrization p_U of type (3.18) or (3.19), terminal control law κ.
1: set $U_C = 0 \in \mathbb{R}^{mN}$
2: **loop**
3:      obtain current state x
4:      find i^* such that $x \in \tilde{\mathcal{D}}_{i^*}$
5:      solve $\mathbf{P}_{p_{U_C},i^*}(x)$, get $\tilde{U}^*$
6:      set $U = p_{U_C,i^*}(x, \tilde{U}^*)$
7:      define $u_{\text{se}}(x) = [I_m \ 0]U$ and apply $u_{\text{se}}(x)$ to the plant
8:      set $U_C = [u_1^\top, \dots, u_{N-1}^\top, \kappa(x_N)^\top]^\top$
9: **end loop**

---

the cost of the candidate input sequence does not have to be compared explicitly to the cost of the newly optimized sequence. Nevertheless, the feasibility, stability and performance results established for the basic semi-explicit scheme carry over to this semi-explicit MPC scheme due to the core argument formulated in the next lemma.

Lemma 3.14. *Let $\tilde{p}_{U_C}$ be a parametrization of type (3.18) and let $\tilde{\tilde{p}}_{U_C}$ be a parametrization of type (3.19). For $U_C \in \mathbb{R}^{mN}$ and $x \in \mathbb{R}^n$ it holds that*

$$U_C \in \operatorname{im}(\tilde{p}_{U_C}(x, \cdot)) \subset \operatorname{im}\left(\tilde{\tilde{p}}_{U_C}(x, \cdot)\right). \quad (3.20)$$

Proof. The first relation follow from $U_C = \tilde{p}_{U_C}(x, [0, 1]^\top)$ and the second relation follows from $\tilde{p}_{U_C}(x, [\tilde{U}^\top, \alpha]^\top) = \tilde{\tilde{p}}_{U_C}(x, [\tilde{U}^\top, 1 - \alpha, \alpha]^\top)$ for $\tilde{U} \in \mathbb{R}^q$, $\alpha \in \mathbb{R}$. □

For the corresponding optimization problems, this implies the following.

Corollary 3.15. *It holds that*

$$\min_{\tilde{U} \in \mathbb{R}^{q+2}} J(x, \tilde{\tilde{p}}_{U_C}(x, \tilde{U})) \leq \min_{\tilde{U} \in \mathbb{R}^{q+1}} J(x, \tilde{p}_{U_C}(x, \tilde{U})) \leq J(x, U_C).$$
$$\text{s.t. } g(x, \tilde{p}_{U_C}(x, \tilde{U})) \leq 0 \qquad \text{s.t. } g(x, \tilde{\tilde{p}}_{U_C}(x, \tilde{U})) \leq 0$$

As a consequence of the latter considerations the following holds.

Corollary 3.16. *Applying Algorithm 5, Inequality (3.13) holds with $\tau = 1$ and relation (3.14) remains valid.*

Trivially, Algorithm 5 is initially feasible for at least the same set of states for which the basic semi-explicit MPC scheme is initially feasible. Together, these facts readily imply that the results established for the basic semi-explicit MPC scheme carry over to the latter algorithm.

Theorem 3.17 (Closed-loop stability). *Consider System (3.1) with corresponding MPC formulation resulting in mpP (3.2). Let Assumptions 2.1, 2.2 and 3.1 regarding the MPC setup hold. Let p be a parametrization of type (3.5) such that the corresponding parametrized mpP $\boldsymbol{P}_p(x)$ is feasible for all $x \in \mathcal{X}_f$ and let Assumption 3.2 hold with respect to p. Let p_U be a corresponding extended parametrization of first or second type applied in Algorithm 5. Then, Algorithm 5 is recursively feasible. It renders the origin of the closed-loop system asymptotically stable and the set $\mathcal{X}_f$ is a subset of the region of attraction of the closed loop.*

Furthermore, Corollary 3.15 implies that Proposition 3.12 holds for Algorithm 5 with $\tau = 1$. As in each time step explicitly a feasible solution is included into the image of the parametrization used, recursive feasibility of the optimization and strong feasibility of the MPC scheme are readily obtained.

Semi-explicit MPC in a dual-mode strategy

Both semi-explicit MPC algorithms (Algorithm 4 and 5) can be modified into a dual-mode strategy. In a dual mode MPC scheme, MPC is used only to drive the system state into a terminal set. Once the state has entered the terminal set, an explicitly known terminal control law is applied directly (Chisci et al., 1996; Scokaert et al., 1999). For the semi-explicit MPC schemes, this would simply mean to additionally check in each time-step if the state has entered the terminal set and, if so, to switch to applying the terminal control law directly. Convergence follows as for the schemes discussed above whereas closed-loop asymptotic stability is contributed by the terminal control law. In this case clearly no requirements regarding continuity of the value function of the parametrized optimization at the origin are needed.

3.5 Examples and evaluation

Having introduced the general semi-explicit MPC strategy including corresponding offline and online algorithms in the previous sections, we next evaluate these results. First, numerical examples illustrate applicability and the potential of the methods. Second, we evaluate the approach conceptually and compare it to existing approaches which simplify the online computations in MPC and share aspects with the semi-explicit MPC approach. For this evaluation, we stick to an abstract level considering the approach and leave more detailed considerations to subsequent chapters.

3.5.1 Numerical examples

Recovering the explicit solution of an mpQP

We use the first example to illustrate that clustering solutions of an mpP via the proposed approach can actually reveal existing structure within the solutions and can yield a parametrization which exactly fits this structure. To this end, we compute parametrizations for a simple mpQP derived from an MPC problem for which the exact explicit solution is available and has moderate complexity.

Example 3.18. *Consider the system*

$$x^+ = \begin{bmatrix} 1 & 1 \\ 0 & 1 \end{bmatrix} x + \begin{bmatrix} 1 \\ 0.5 \end{bmatrix} u$$

with constraints $\|x\|_\infty \leq 4$, $|u| \leq 1$. In the MPC problem we use the stage cost $\ell(x, u) = \|x\|_2^2 + u^2$, the terminal controller and quadratic terminal cost according to the corresponding unconstrained LQR setup and the terminal set corresponding to the associated maximal output admissible set (Gilbert and Tan, 1991). The prediction horizon is set to $N = 8$. The solution of the resulting mpQP consists of a piecewise affine optimizer defined on 25 critical regions. Having this fact in mind, we computed parametrizations fixing $\phi(x) = [x^\top, 1]^\top$ such that the parametrization has the structure $p_i(x, \tilde{U}) = M_i \tilde{U} + \mathcal{K}_i x + a_i$ and, thus, covers for $q = 0$, $K \geq 25$ the explicit solution of the mpQP.

We applied the approach to compute parametrizations (Algorithm 3) using training states sampled from a regular grid of 100×100 points which exactly covers the feasible set of the mpQP and tested different combinations of values for K and q. In particular for each value of $q \in \{0, 1, \ldots, 7\}$ we determined the lowest value of K such that the training data was perfectly contained in the image of the parametrization, i.e., $J_C = 0$ was achieved in the clustering algorithm. Using $q = 0$, one can hope to actually recover the explicit solution and considering $q = 7$ as largest value is motivated by the fact that $q = 8$ already corresponds to the original non-parametrized version. The results are illustrated in Figure 3.1 where only the combinations at the Pareto frontier are shown[2] (back dots) and in addition the non-parametrized version is marked (gray dot). The K-q-combinations for which the image of the parametrization exactly contains the training data illustrates a tradeoff of both hyperparameters. Considering that the explicit mpQP solution comprises 25 critical regions, it looks as if the parametrization computed with ($q = 0$, $K = 25$) is this explicit solution up to the definition of the critical regions.

---

[2] The cases $q \in \{5, 6, 7\}$ are omitted in the figure as for these values of q still $K \geq 2$ was required to achieve $J_C = 0$.

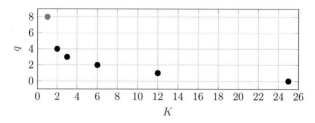

Figure 3.1: Several combinations of K and q for which training data was perfectly contained in image of parametrization for Example 3.18.

It is much more interesting and insightful to consider the clustered points in state space and compare the clusters to the critical regions of the explicit mpQP solution. Figure 3.2 shows for the combinations ($q = 0$, $K = 25$), ($q = 1$, $K = 12$) and ($q = 2$, $K = 6$) the clustered states colored according to cluster membership together with the boundaries of the critical regions shown in black.

For the case ($q = 0$, $K = 25$), the critical regions are actually perfectly resembled by the clustered data. Each critical region contains states from exactly one cluster and no cluster spreads over more than one critical region. In the case ($q = 1$, $K = 12$), clusters mainly spread over two critical regions which share a common facet. In two cases states in two critical regions which are located point-symmetrically to each other with respect to the origin are assigned to the same cluster (above left and below right from the central region). It can be shown analytically that in both cases (i.e. merging adjacent and symmetrically located critical regions) one degree of freedom ($q = 1$) is sufficient to cover the solution of both critical regions using a common part p_i of the parametrization.[3] In the case ($q = 2$, $K = 6$), both aforementioned strategies of merging critical regions appear combined. States in pairwise adjacent critical regions and corresponding point-symmetrically located regions are assigned to the same cluster. In addition, there seems to be some ambiguity which causes several clusters to spread into the central critical region and into regions above left from the central region.

Summarizing, in this example the method to compute the parametrization proved to be capable of perfectly recovering existing non-trivial structure in mpP solutions. Via a suitable selection of the hyperparameters K and q, a gradual transition was possible from recovering the explicit solution to finding parametrizations of different complexity which still contain the exact solution perfectly in their image.

Approximating a complex solution

In the next example, we apply the proposed algorithms to an mpP which has a solution of high complexity. The goal here is to illustrate the general method, to demonstrate the effect of the hyperparameters and to show the capability to approximate complex mpP solutions via the parametrization such that simplified MPC schemes can be formulated

---

[3] In fact the initialization of the clustering influenced the ratio with which both merging strategies were observed and using a parametrization without constant part (i.e. enforcing $a_i = 0$) only admits the symmetric merging.

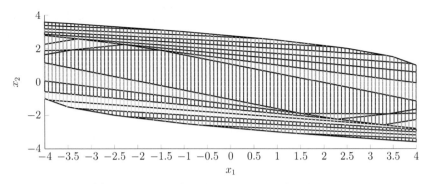

(a) $K = 25$, $q = 0$. Critical regions recovered as clusters.

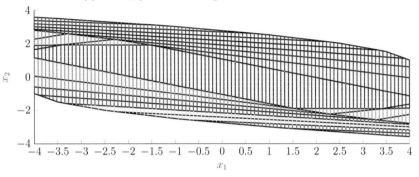

(b) $K = 12$, $q = 1$. Mainly adjacent critical regions merged into common cluster.

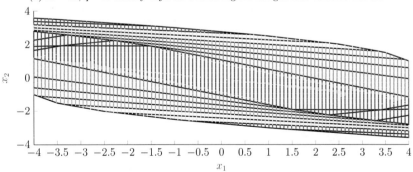

(c) $K = 6$, $q = 2$. Adjacent and point-symmetrically located critical regions merged into common cluster.

Figure 3.2: Clustered states in state space and bounds of critical regions of explicit solution for Example 3.18.

based thereon. Furthermore, the correlation of i) approximation accuracy of the training data, ii) open-loop performance and iii) closed-loop performance will be illustrated.

Example 3.19. *Again we consider an mpQP related to an MPC problem addressing control of a linear system $x^+ = Ax + Bu$, in this case with $x \in \mathbb{R}^6$, $u \in \mathbb{R}^3$. The matrices A and B were generated randomly with the absolute values of the eigenvalues of the dynamic matrix A confined to the interval $[0.9, 0.95]$ and they are given by*

$$A = \begin{bmatrix} 0.5800 & -0.1401 & 0.3541 & 0.8212 & 0.1451 & 0.1123 \\ 0.8062 & 0.4313 & -0.7333 & 0.7489 & 0.7753 & 0.6838 \\ -1.1217 & -0.1126 & 0.0692 & -0.0010 & 0.4274 & -0.5895 \\ 0.3371 & -0.3189 & -0.8150 & -0.0776 & -0.1821 & 0.8313 \\ 0.0185 & -0.7605 & 0.7403 & 0.9691 & -0.7605 & -0.0319 \\ 0.8574 & -0.7150 & -0.2624 & 0.8189 & -0.3279 & -0.4332 \end{bmatrix}$$

and

$$B = \begin{bmatrix} 0.4055 & 0.9137 & -0.4953 \\ 0.0931 & 0.0452 & 0.7515 \\ -0.1102 & 0.7603 & 0.4746 \\ 0.3891 & -0.6541 & -0.7270 \\ 0.2426 & 0.9595 & -0.9765 \\ 0.5896 & -0.4571 & 0.7878 \end{bmatrix}.$$

In particular, the autonomous system is stable. The stage cost $\ell(x, u) = \|x\|_2^2 + \|u\|_2^2$ is used and the input constraints $\|u\|_\infty \le 1$ are enforced. For the MPC setup the terminal cost function $V_T(x) = x^\top P x$ is used with $P > 0$ satisfying the Lyapunov equation

$$A^\top P A - P = Q$$

with $Q = I$ and no terminal constraint ($\mathbb{X}_T = \mathbb{R}^6$) is enforced. The prediction horizon is set to $N = 10$. For this setup, the stability implying Assumption 2.3 is satisfied with terminal control law $\kappa(x) \equiv 0$. Due to the chosen setup, $u_0 = u_1 = \cdots = u_9 = 0$ is an admissible solution of the finite horizon optimal control problem in the MPC formulation and, thus, $U = 0$ is an admissible solution of the resulting mpQP. Note that the original QP comprises a 30-dimensional decision variable and the explicit solution of the mpQP comprises more than 15 000 critical regions.

We computed several parametrizations applying Algorithm 3 with $s = 4^6$ training states sampled regularly from the box $[-4, 4]^6$. The hyperparameters $K \in \{1, 5\}$ and $\phi \in \{\phi_1, \phi_2, \phi_3\}$, with $\phi_1 \equiv 0$, $\phi_2(x) = x$, $\phi_3(x) = [1, x^\top, \sin(\pi/5x)]^\top$ were used[4] where the sine is applied element-wise. The sets $\mathcal{D}_i$ were defined by assigning each point to the same cluster as the closest training state. We evaluate the obtained results in three steps.

1. *First, we analyze the capability of the different parametrizations to approximate the training data used. This is done by comparing the achieved objective function value J_C in the clustering procedure for the different hyperparameter values. The results are summarized in Table 3.1. As is to be expected, increasing the complexity of the parametrization by increasing the value of K or by choosing more general functions ϕ improves the approximation accuracy of the training data.*

---

[4] The scaling of the argument in the sine term is motivated by the box of initial conditions considered here. In applications incorporating sinusoidal terms might unnecessarily complicate matters and we use them here only for illustration.

2. *Second, we evaluate how well the computed parametrizations are suitable to approximate solutions of the mpQP for states different from the training states used. In order to guarantee feasibility of the parametrized optimization problems, 0 is included into the image of the parametrization by transforming the parametrizations into parametrizations of type (3.18), but setting $U_C = 0$ therein. See the first row of Table 3.2 for the resulting parametrizations. Note that the value of q is chosen such that in all cases the decision variables have the same dimension. The different parametrized mpQPs are each solved for the same set of 10 000 states sampled randomly from the box $[-5, 5]^6$ and the average of the relative cost increase*

$$\mathcal{P}^{ol}(x) = \frac{V_p^*(x)}{V^*(x)} - 1 \qquad (3.21)$$

over all states is computed. All values are given in Table 3.2.

The same trends as in the first stage of the evaluation are apparent. Making the single parts of the parametrization state-dependent improves the approximation largely. An interpretation of this observations is that the parametrizations actually generalize approximation of mpQP solutions for training states to approximation of general mpQP solutions and using more complex parametrizations actually pays off in better approximation of solutions. If the parametrizations largely over-fitted the training data used, using a more complex parametrization would generally not translate into better approximation of unseen data. The fact that approximation accuracy is high also for unseen data, hints that over-fitting did not happen.

3. *Third, the obtained parametrizations are employed in a semi-explicit MPC scheme and are evaluated via closed-loop simulations. To this end, another set of 10 000 states is sampled randomly from the box $[-5, 5]^6$ and used as initial conditions applying Algorithm 5. Here, parametrizations of type (3.19) were used, initially setting $U_C = 0$ and then using the actual candidate solution U_C. All closed-loop simulations are executed until the state has entered an ellipsoidal set centered at the origin, which is invariant under the optimal input and the optimal input satisfies the constraints for all states in the ellipsoid. The cost-to-go from this state on was then incorporated via the terminal cost. Suboptimality of the semi-explicit MPC scheme is then evaluated according to*

$$\mathcal{P}(x) = \frac{J_{seMPC}^{cl}(x)}{J_{nom}^{cl}(x)} - 1 \qquad (3.22)$$

again taking the average over all initial conditions. The numerical values observed are given in Table 3.3. As in the previous steps of the evaluation, performance clearly improves with growing complexity of the parametrizations used and in fact there seems to be a close correlation of J_C, $\mathcal{P}^{ol}$ and $\mathcal{P}$. Interesting to note is that the closed-loop suboptimality is considerably lower than the open-loop suboptimality. This is probably due to the fact that the input sequence used is improved along closed-loop trajectories over the initial open-loop prediction as repeatedly new solutions were computed and the old solution was included into the image of the parametrization.

Note that in this case the explicit solution of the mpQP is highly complex and consists of more than 15 000 critical regions. So recovering this exact solution is not possible with

Table 3.1: Value of the clustering objective function J_C observed for different combinations of hyperparameters for Example 3.19.

	$\phi(x) \equiv 0,\ q = 6$	$\phi(x) = x,\ q = 5$	$\phi(x) = [1; x^\top; \sin(\pi/5x)]^\top,\ q = 5$
$K = 1$	6457	3587	3566
$K = 5$	3639	1500	1464

Table 3.2: Average of relative open-loop cost increase $\mathcal{P}^{\mathrm{ol}}(x)$ over 10 000 randomly sampled states observed for parametrizations of different complexity for Example 3.19.

	$\phi(x) \equiv 0,\ q = 6$ $\tilde{\tilde{p}}_i(x, \tilde{U}) = M_i \tilde{U}$	$\phi(x) = x,\ q = 5$ $\tilde{\tilde{p}}_i(x, \tilde{U}) = [M_i \ \ \mathcal{K}_i x] \tilde{U}$	$\phi(x) = [1; x^\top; \sin(\pi/5x)]^\top,\ q = 5$ $\tilde{\tilde{p}}_i(x, \tilde{U}) = [M_i \ \ \mathcal{K}_i x + \tilde{\mathcal{K}}_i \sin(\pi/5x) + a_i] \tilde{U}$
$K = 1$	18.5%	7.0%	6.9%
$K = 5$	15.0%	5.2%	4.6%

Table 3.3: Average of closed-loop suboptimality $\mathcal{P}(x)$ over 10 000 randomly sampled initial conditions observed for parametrizations of different complexity for Example 3.19.

	$\phi_1(x) \equiv 0,\ q = 6$ $\tilde{\tilde{p}}_i(x, \tilde{U}) = [M_i \ \ U_C] \tilde{U}$	$\phi(x) = x,\ q = 5$ $\tilde{\tilde{p}}_i(x, \tilde{U}) = [M_i \ \ \mathcal{K}_i x \ \ U_C] \tilde{U}$	$\phi(x) = [1; x^\top; \sin(\pi/5x)]^\top,\ q = 5$ $\tilde{\tilde{p}}_i(x, \tilde{U}) = $ $[M_i \ \ \mathcal{K}_i x + \tilde{\mathcal{K}}_i \sin(\pi/5x) + a_i \ \ U_C] \tilde{U}$
$K = 1$	7.5%	2.0%	1.9%
$K = 5$	4.5%	1.6%	1.4%

reasonable effort nor desirable from an application point of view. Nevertheless, all offline computations applying the proposed methods were well tractable for this example. The computation times required for the clustering procedure were, depending on the hyperparameters used, 0.6 s up to 19.9 s running it from 20 different initial conditions.

Summarizing, in this example parametrizations were computed and applied for an MPC problem where the exact mpQP solution is rather complex. Parametrizations using different hyperparameters were used to approximate optimal solutions with different accuracy. A close correlation of the achieved clustering objective function value J_C, the open-loop performance $\mathcal{P}^{ol}$ and the closed-loop control performance $\mathcal{P}$ was observed.

3.5.2 Evaluation and possible extensions of the clustering algorithm

In the following, we take a closer look at the characteristics of the clustering algorithm employed. As discussed above, the proposed algorithm follows the paradigm of iterative clustering algorithms. Various other classes of subspace clustering algorithms have been reported in the literature which follow different strategies, see (Vidal, 2011) for an overview. One example is the so-called sparse subspace clustering (SSC) algorithm (Elhamifar and Vidal, 2013) which is based on computing a sparse representation of the data points and using this sparse representation to identify clusters and corresponding subspaces. SSC has been applied successfully to find parametrizations for MPC, see (Alber, 2013). Nevertheless, the iterative method employed here has a number of features which make it advantageous over other existing subspace clustering methods including SSC.

Due to the fact that its procedure is composed of very simple elementary steps, it is amenable to various extensions into several directions. Without going into details, we next briefly indicate the most important ones of which some have been applied or will be applied below in this thesis.

- The first extension over classical pure subspace clustering algorithms was to establish in addition a linear correlation within each cluster of the state x and the input sequence U as reported in (Goebel and Allgöwer, 2014a). Here this technique was further extended to establishing a more general linear correlation of $\phi(x)$ and U. This follows similar ideas as using the "kernel trick" in machine learning, where instead of working in a linear fashion directly on the data, the problem is (implicitly) transferred to a higher dimensional space where one can again work in a linear fashion but exploit more degrees of freedom, see for example (Hofmann et al., 2008).

- A straight forward extension is to weight the single elements of the U-vectors differently, i.e., to make approximation of some elements in U by the parametrization more important. This can simply be achieved via executing weighted least squares and low rank approximations in the clustering algorithm and taking this weighting into account when assigning data to the clusters. In our context this is of paramount relevance in order to put a higher weight on the first elements of the predicted input sequence. This mimics in some sense a move-blocking scheme which distributes more flexibility towards the beginning of a predicted sequence and accounts for the fact that generally only the first part of a predicted sequence will be applied to the plant.

Beyond that, if input signals of the system to be controlled are in different orders of magnitude, weighting (or scaling) them to ensure that they are all taken into account equally in the clustering is a reasonable step to take.

- A similar extension is to weight the single training states used differently, for example to increase approximation accuracy in areas of the state space which are more important because they are visited more frequently by closed loop trajectories. Weighting of the training data will also be used to refine parametrizations for nonlinear systems in Subsection 6.3.3 of this thesis. The weighting can again be achieved via incorporating the weighting in the respective steps of the clustering algorithm.

- Beyond that, the clustering can be extended by adding terms to the objective function used or by enforcing additional constraints. An additional term in the objective function will be used below in this thesis to foster compactly shaped clusters in state space for nonlinear problems (Subsection 6.3.2) whereas additional constraints regarding cluster membership of the data points will be used in a linear setting to assign whole simplices to the same cluster and thereby obtain suitable sets $\hat{\mathcal{D}}_i$ conveniently (Subsection 4.3.2).

- Further extensions are possible but are not considered in this thesis. Inspired by available versions of k-means clustering, they could for example cover adaptation of the clustering to different data types as discrete valued or binary data, using other distance measures, adding regularizing or sparsity promoting terms etc.

One feature of the class of clustering algorithms employed here is that the hyperparameters K and q (and also ϕ) have to be chosen a priori by the user. From a pure data mining perspective, this is clearly a disadvantage as, ideally, one would like to have the algorithm find the structure of the data under consideration and in doing so also identify these values. In our application, this feature is in fact an advantage as it allows to adjust the characteristics of the obtained parametrization. The first example above (Example 3.18) actually showed that there are cases where several choices on the K-q-grid yield parametrizations which approximate the data equally well (and in fact ideally) and so the proposed algorithm equips the user with the possibility to choose one of these options. The possibility to choose the hyperparameters becomes even more important in order to adjust complexity versus approximation accuracy.

Finally, a paramount advantage of the proposed clustering scheme is its simplicity which makes it well scalable and applicable to large and high dimensional data sets. In (Agarwal and Mustafa, 2004) a similar algorithm was applied to problem sets consisting of 50 000 data points which are up to 300-dimensional and still running times of only few minutes were observed. Regarding scalability of the whole semi-explicit MPC approach, this is a very strong argument.

3.5.3 Discussion of the semi-explicit MPC approach and comparison to related approaches

We next evaluate the semi-explicit MPC approach on a conceptual level. A more detailed evaluation is postponed to the next chapter where readily applicable schemes for linear systems are developed and tested. Considering the employed methods, the proposed semi-explicit MPC approach has three key features: First, it exploits the parameter-dependence of the optimization problem to be solved online without computing or using a complete explicitly solution of the problem. Second, it employs parametrizations in order to simplify the online optimization. Third, the method's ingredients are computed data-based. Hence, we compare the proposed method to existing methods which share at least one of these aspects. The relations of these results among each other and to the proposed method are illustrated graphically in Figure 3.3.

Results which employ parametrizations typically do not exploit state-dependence of the optimization problem they are applied for. An exception in this sense are interpolation based schemes, which could be seen as state-dependent parametrizations. As stated in (Khan et al., 2014), when applying more elaborate parametrizations in MPC, one is typically interested in improving the tradeoff between the region of attraction, the control performance and the online computational burden. As semi-explicit MPC introduces the possibility to store state-dependent information and exploit it online, it adds another aspect to the tradeoff. This aspect is adjustable via the chosen value for K and q and via the complexity of the function ϕ. In Example 3.18 this tradeoff was illustrated, as parametrizations with different "degree of explicitness" all recovered the exact solution.

Data-based approaches typically aim at finding an explicit map from the state to the corresponding optimal input. As semi-explicit MPC extracts and uses only structural information about this mapping, it can generally be expected to be simpler offline and require less data to be stored and made available online. In contrast to most data-based approaches, the proposed method is capable of revealing actually *existing* structure in the data (see Example 3.18).

Results which exploit parameter-dependence of the optimization are rather diverse, but they have in common that rarely any of them works based on data. For example, in (Ferreau et al., 2008) an online active set strategy is formulated which founds on using the previous solution and insights into the structure of active sets in state space to find the next active set. An innovative type of results making use of the parameter-dependence of the optimization exploits information on the inactivity of constraints. In (Jost and Mönnigmann, 2013a,b) state dependent information on inactivity of constraints is generated offline and used online, whereas in (Jost et al., 2015, 2017) information depending on the value of a Lyapunov function is generated which can be used to conclude on inactive constraints online. Removing constraints from the optimization problem which are known to be inactive clearly simplifies its numerical solution. The latter results are similar to semi-explicit MPC as they replace the optimization problem by a simplified one based on a priori offline prepared data.

Semi-explicit MPC as proposed in this thesis is unique as it exploits parameter-dependence of the optimization, employs a parametrization and computes its ingredients data-based. Due to this combination, it possesses a number of advantages over existing approaches. A main advantage of the general semi-explicit MPC approach is its applicability to broad system and problem classes. This is due to the fact that it works data-based, employs a very general simplification of the online optimization (i.e. reduction of the dimension of the decision variable via a parametrization) and exploits for this purpose parameter-dependence of the optimizers in a very general sense. In this thesis, we focus on setpoint stabilizing MPC schemes for nominal systems. Nevertheless, the general approach is by no means restricted to this class of control problems and the core steps of the approach work completely problem independent. The main remaining question is how to establish feasibility guarantees for the parametrized optimization in different situations. In the sequel of this thesis, different solution strategies for this problem will be presented.

A second advantage of the general proposed approach is that its algorithms are composed of rather simple elementary steps and thus it scales relatively well in the problem dimension. In decomposing the goal of finding a suitable parametrization during the offline phase into simple steps, the approach relies on a "cascade of objectives". In a first step, a parametrization is computed such that the optimal solution of the considered mpP for a finite number of training states is approximated well. In a second step, one relies on that this parametrization is also suitable to approximate solutions for states which are close to the considered states but which have not been considered explicitly before. In a third stage, high approximation accuracy for general solutions is expected to carry over to high control performance if applied in the semi-explicit MPC scheme. Whereas the relation of these objectives clearly is not monotonic[5], they generally correlate well as is observed in numerical examples and as can be reasoned theoretically. Theoretically, the first connection is justified by regularity properties of the mpP solution and the second connection is verified via the upper bound on the closed-loop cost established in Proposition 3.12. The three objectives and their connections are illustrated in Example 3.19: Performance measures for all three stages are given in Table 3.1, Table 3.2 and Table 3.3 and clearly the performance measures correlate well. Feasibility issues of the parametrized optimization have not yet been addressed explicitly so far. When this aspect is incorporated and considered in the chapters below, furthermore an implicit assumption on the connection of approximation accuracy and feasibility is made.

3.6 Summary

This chapter was dedicated to the introduction of our general semi-explicit MPC scheme as well as the basic ingredients of its offline and online algorithms. The fundamental idea of the semi-explicit MPC scheme is to first compute in an offline part state-dependent parametrizations for the solutions of the MPC optimization problem via a data-based approach. In the second online part of the semi-explicit MPC scheme, these parametrizations are applied to simplify and accelerate numerical solution of the optimization.

In this chapter we first proposed the parametrization together with a suitable procedure to compute them for a given problem. This procedure applies a tailored subspace clustering algorithm which was contributed as well. We verified that parametric programs maintain beneficial properties if the proposed parametrizations are applied therein. Based on these parametrized optimization problems, the online part of the semi-explicit scheme was then formulated. For the resulting closed loop, asymptotic stability and recursive feasibility results and a bound on closed-loop performance were formulated. Finally, two numerical examples were presented. The first example illustrated that the method can reveal existing

---

[5]Non-monotonicity of the relations is immediately clear from the fact that average closed-loop control performance over randomly sampled initial conditions is affected stronger by open-loop performance of states close to the origin as this region is visited by more of the closed-loop state trajectories and, similarly, quality of predicted inputs towards the beginning of the prediction horizon are more important for control performance as they are more likely to be actually applied to the plant. Nevertheless, all predicted inputs and all states affect the respective performance measures equally.

exploit parameter-dependence
for online optimization

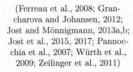

(Ferreau et al., 2008; Grancharova and Johansen, 2012; Jost and Mönnigmann, 2013a,b; Jost et al., 2015, 2017; Pannocchia et al., 2007; Würth et al., 2009; Zeilinger et al., 2011)

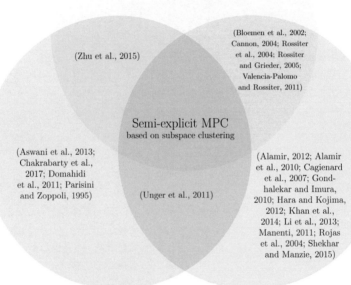

(Zhu et al., 2015)

(Bloemen et al., 2002; Cannon, 2004; Rossiter et al., 2004; Rossiter and Grieder, 2005; Valencia-Palomo and Rossiter, 2011)

Semi-explicit MPC
based on subspace clustering

(Aswani et al., 2013; Chakrabarty et al., 2017; Domahidi et al., 2011; Parisini and Zoppoli, 1995)

(Unger et al., 2011)

(Alamir, 2012; Alamir et al., 2010; Cagienard et al., 2007; Gondhalekar and Imura, 2010; Hara and Kojima, 2012; Khan et al., 2014; Li et al., 2013; Manenti, 2011; Rojas et al., 2004; Shekhar and Manzie, 2015)

data-based approaches   parametrizations

Figure 3.3: Relation of the semi-explicit MPC approach to existing approaches. Semi-explicit MPC as proposed in this thesis is unique in combining all three of the shown aspects. Yet, it is much more innovative than a mere combination of three established results into a new one would be. The chosen categories are general and abstract fields in their own, and the proposed results even contribute to the available results within these categories. The proposed results go beyond available parametrization based results, they differ considerably from most existing data-based approaches and they contribute a new way to exploit parameter-dependence for the online optimization.

structure in the solutions of an mpQP and contains the explicit solution as a special case. In the second numerical example, a close correlation of different performance measures from approximation of training data up to closed-loop control performance was shown. Both examples illustrated the possible tradeoff of offline preparations versus required online computational load. The question of how to find parametrizations which ensure that the parametrized optimization problems are guaranteed to be feasible for a certain given set of states was not addressed in this chapter and is a main subject of subsequent chapters instead.

Chapter 4

Semi-explicit MPC for linear systems

In the previous chapter, we introduced our general semi-explicit MPC approach. A number of generic results has been established which, just as the method itself, are applicable to large classes of problems as no specific properties of the underlying system, constraints or cost functions were assumed nor exploited. Our goal in the current chapter is to refine and extend these generic results by restricting attention to the class of linear systems with affine constraints. Tailoring the general method to this system class, readily applicable offline and online algorithms for the semi-explicit MPC scheme will be developed and presented.

This chapter is partly based on (Goebel and Allgöwer, 2013, 2014a, 2017a,b).

4.1 Introduction and problem formulation

In this chapter, we address control of a linear discrete-time system

$$x_{k+1} = Ax_k + Bu_k, \ \ k \geq 0 \tag{4.1}$$

with given initial condition x_0 and state and input constraints

$$x_k \in \mathbb{X} \subseteq \mathbb{R}^n, \ \ u_k \in \mathbb{U} \subset \mathbb{R}^m$$

in form of polytopic sets, cf. (2.6) and (2.7). All specifications and general assumptions regarding the MPC setup used throughout the previous chapter remain valid: A setpoint stabilizing MPC scheme is considered, which comprises a terminal cost and terminal constraints which together with a known terminal control law fulfill the stability implying Assumption 2.3, i.e., we keep Assumption 3.1. Also Assumptions 2.1 and 2.2 remain valid. In addition, also the terminal set $\mathbb{X}_T$ is assumed to be a polytope, cf. (2.10). Recall that under these assumptions the general finite horizon open-loop optimal control problem (2.2) can be reformulated as

$$\begin{aligned} &\min_U J(x, U) \\ &\text{s.t.} \ \ GU \leq W + Ex. \end{aligned} \tag{4.2}$$

If additionally the stage cost and the terminal cost are assumed to be quadratic functions, the objective function of the latter mpP has the form

$$J(x, U) = U^\top HU + x^\top FU + x^\top Yx. \tag{4.3}$$

In the following, we first examine properties of parametrized multi-parametric programs which correspond to such linear MPC problems. Second, for this problem class three approaches are proposed to compute parametrizations which are guaranteed to result in a feasible parametrized optimization problem for a given polytopic set of states. Then, properties of the resulting online MPC scheme will be examined and an offline stability test will be formulated, which allows to employ a simplified online scheme and still guarantee asymptotic stability of the closed loop. Finally, the results for linear systems are evaluated both via numerical examples and theoretically, and we will compare them to existing MPC approaches applicable to the considered problem class.

4.2 The parametrizations applied to linear MPC problems

The crucial benefit of restricting the focus to linear systems with polytopic constraints for the proposed method is the resulting structural simplicity of the constraints in the mpP (4.2). Most importantly, the constraints are jointly convex in the state x and in the decision variable U. Furthermore, the constraints define a polytope of admissible (x, U) combinations in the $x \times U$ space which, projected onto the state space, results in a set of feasible x values which again is a polytope. In particular, this means that the set of states for which mpP (4.2) is feasible is a convex polytope. If parametrizations are employed, they should be chosen in a way such that convexity and the polytopic nature of the feasible set is maintained. Thus, the parametrizations considered in this chapter are structurally restricted with respect to the general parametrizations introduced in the previous chapter.

Definition 4.1. *A parametrization according to Definition 3.1 for which $\phi(x) = [x^\top, 1]^\top$ holds is called a* parametrization for linear systems.

The most general parametrizations considered in the current chapter are of this type and for some results we will use $\phi(x) = x$. Let $\mathcal{K}_i \in \mathbb{R}^{mN \times n}$, $a_i \in \mathbb{R}^{mN}$ and let N_i be as defined in the general parametrization. Identifying the relation $[\mathcal{K}_i \ a_i] = N_i$, a parametrization for linear systems defines a mapping $p : \mathbb{R}^n \times \mathbb{R}^q \to \mathbb{R}^{mN}$, $(x, \tilde{U}) \mapsto p(x, \tilde{U})$ with

$$p(x, \tilde{U}) = \begin{cases} p_1(x, \tilde{U}) = M_1 \tilde{U} + \mathcal{K}_1 x + a_1 & \text{if } x \in \mathcal{D}_1 \\ \vdots \qquad \qquad \vdots & \vdots \\ p_K(x, \tilde{U}) = M_K \tilde{U} + \mathcal{K}_K x + a_K & \text{if } x \in \mathcal{D}_K \end{cases} \tag{4.4}$$

where $M_i \in \mathbb{R}^{mN \times q}$, $i = 1, \ldots, K$ and $\mathcal{D}_i \subset \mathbb{R}^n$ and $q < mN$ holds. Throughout this chapter, we make the non-restrictive assumption that the matrices M_i have full column rank. As the matrices have less columns than rows ($q < mN$), matrices which violate this assumption can always be adjusted to meet the assumption by exchanging some columns and thereby enlarging the image of the parametrization. Using a parametrization of the structure (4.4), it is immediately clear, that the constraints of the parametrized multi-parametric program obtained for a linear system possess the same structure as the constraints of the non-parametrized one.

Lemma 4.2. *A parametrization for linear systems applied in mpP (4.2) results for $x \in \mathcal{D}_i$ in a parametrized mpP*

$$\min_{\tilde{U}} J(x, p_i(x, \tilde{U}))$$
$$s.t. \quad \tilde{G}_i \tilde{U} \leq \tilde{W}_i + \tilde{E}_i x \tag{4.5}$$

for matrices $\tilde{G}_i = GM_i$, $\tilde{W}_i = W - Ga_i$ and $\tilde{E}_i = E - GK_i$. In particular, the set of states for which (4.5) is feasible is a convex polytope.

As these are still affine constraints, identifying redundant ones among them is possible in a straight forward manner via solving one linear program per constraint, see e.g. (Borrelli et al., 2015). A result similar to the latter one holds regarding quadratic objective functions.

Lemma 4.3. *A parametrization for linear systems applied in the mpP (4.2) with objective function (4.3) results for $x \in \mathcal{D}_i$ in a parametrized mpP*

$$
\begin{aligned}
\min_{\tilde{U}} \ & \tilde{U}^\top M_i^\top H M_i \tilde{U} + \left\{ x^\top \left(F M_i + 2\mathcal{K}_i^\top H M_i \right) + 2a_i^\top H M_i \right\} \tilde{U} + k(x) \\
\text{s.t.} \ & \tilde{G}_i \tilde{U} \le \tilde{W}_i + \tilde{E}_i x
\end{aligned}
\tag{4.6}
$$

where

$$
k(x) = a_i^\top H a_i + \left\{ x^\top \left(Y + \mathcal{K}_i^\top H K_i + F \mathcal{K}_i \right) + a_i \left(2H\mathcal{K}_i + F^\top \right) \right\} x
$$

and matrices $\tilde{G}_i$, $\tilde{W}_i$ and $\tilde{E}_i$ are defined as in (4.5). In particular, for fixed state x, (4.6) is a convex quadratic program. If (4.3) is jointly convex in x and U, the objective function of (4.6) is jointly convex in x and $\tilde{U}$.

Proof. The structure (4.6) results again directly from algebraic manipulations, the second claim follows from H being positive definite. To show joint convexity of (4.6) in x and $\tilde{U}$, assume (4.3) is jointly convex in x and U and consider it as a convex function of $[x^\top, U^\top]^\top$. The mapping $[x^\top, \tilde{U}^\top]^\top \mapsto [x^\top, p_i(x, \tilde{U})^\top]^\top$ is affine. So the composition of (4.3) and this affine mapping is again convex. □

Considering $\tilde{x} = [x^\top \ 1]^\top$ as parameter, (4.6) is an mpQP according to Definition 2.4. As joint convexity is maintained using the parametrization, Theorem 2.8 is applicable to the parametrized mpQP if it is applicable to the non-parametrized mpQP.

Considering the single parts $\mathbf{P}_{p_i}(x)$ of the parametrized mpQP jointly, some convexity properties are lost. Assuming that the feasible sets of the single parts are overlapping, the sets of states where one part $\mathbf{P}_{p_i}(x)$ of the parametrized multi-parametric quadratic program yields the best solution $(\mathcal{B}_i = \{x | V_{pi}^*(x) \le V_{pj}^*(x), \ j = 1, \ldots, K\})$ is in general neither convex nor a polytope. Recalling that the optimal value of an mpQP is a piecewise quadratic function, this claim is immediately clear from the fact that such set $\mathcal{B}_i$ is defined via quadratic inequalities resulting from comparing such piecewise quadratic functions. The feasible set of the overall parametrized optimization $\mathbf{P}_p(x)$ is the union of the single feasible sets and, thus, generally not convex either.

The case of an mpLP is obtained by setting $H = 0$ in (4.6). This immediately shows that parametrizing an mpLP, again a parametrized program with linear objective function results.

Finally, we note that the structure of the parametrization considered in this chapter is compatible with two frequently applied techniques when using the condensed formulation of the finite horizon open-loop optimal control problem in linear MPC: On the one hand, it is common practice to pre-stabilize unstable systems in order to improve numerics for the condensed problem formulation (Rossiter et al., 1998). This basically means to treat a stabilized closed-loop system instead of the original open-loop system and consider inputs to the stabilized system which have been transformed correspondingly.

On the other hand, in cases with quadratic objective functions completing the square in the mpQP, i.e. changing coordinates such that the mpQP has a purely quadratic objective function, can simplify some considerations. Both techniques boil down to a transformation of the original decision variable U to a new decision variable Z via a mapping $U \mapsto Z = T_1 U + T_2 x$ where $T_1 \in \mathbb{R}^{mN \times mN}$ is regular and $T_2 \in \mathbb{R}^{mN \times n}$ holds. Clearly, the constraints and objective function have to be transformed accordingly. It is straight forward to transform a parametrization for linear systems in between such representations: $U = M_i \tilde{U} + \mathcal{K}_i x + a_i, \rightsquigarrow Z = T_1 M_i \tilde{U} + (T_1 \mathcal{K}_i + T_2) x + T_1 a_i$. As such transformation leaves the parametrization structurally unchanged, one is free to change to a different representation of the mpP once the parametrization has been determined.

4.3 Computing feasible parametrizations: The offline algorithm

Next, the offline part of the semi-explicit MPC scheme for linear systems is addressed more in detail. Therein, it is the goal to find a parametrization such that applying the parametrization in an mpP feasibility is achieved and can be guaranteed for a given predefined set of states. Recall that we call a parametrization feasible if for a given set $\mathcal{X}_f$ there exist sets $\{\hat{\mathcal{D}}_1, \ldots, \hat{\mathcal{D}}_K\}$ with

$$\mathcal{X}_f \subseteq \cup_{i=1}^{K} \hat{\mathcal{D}}_i \tag{4.7}$$

and

$$\mathbf{P}_{p_i}(x) \text{ is feasible for all } x \in \hat{\mathcal{D}}_i, \tag{4.8}$$

cf. Definition 3.3. In most existing results on parametrizing the decision variable in linear MPC, the parametrization is fixed first and the feasible set is obtained as a result, but, the feasible set of the parametrized optimization problem is not influenced directly. First of all, not a given predefined set of states can be rendered feasible for the parametrized optimization. Generally, this is a reasonable approach which keeps the offline procedure simple and typically causes only little losses at the edges of the original feasible set. Using a linear or affine mapping as parametrization, "holes" cannot appear in the interior of the set of feasible states due to convexity. For the proposed parametrization, nevertheless, the situation is different and this approach is not applicable. Due to the piecewise definition of the parametrizations, the overall feasible set is obtained as the union of several feasible sets of the single parametrized optimization problems $\mathbf{P}_{p_i}(x)$. Loosing feasibility at the edge of one of these patches might mean to loose feasibility in the interior of the overall feasible set. Clearly, having "holes" in the interior of the overall set of feasible states, on the other hand, is impracticable and, thus, unacceptable. Hence, for the proposed approach it is crucial to be able to determine each part of the parametrization p_i such that feasibility of the corresponding parametrized optimization can be guaranteed for a predefined set of states $\hat{\mathcal{D}}_i$. Considerable effort will be put into formulating offline procedures which achieve this task.

Summarizing, the problem we consider in this section is as follows. Given an mpP (4.2) and a set $\mathcal{X}_f \subseteq \mathbb{X}_N$, find a parametrization which is feasible on $\mathcal{X}_f$. Therein, we require $\mathcal{X}_f$ to be a full-dimensional bounded polytope with the origin in its interior and, clearly, for the problem to be feasible the set $\mathcal{X}_f$ has to be contained in the feasible set $\mathbb{X}_N$ of the original mpP. We generally call $\mathcal{X}_f$ the *desired feasible set*. In the sequel, three different algorithms to solve the latter problem will be presented. All three approaches are generally independent of the stage cost and the terminal cost function used in the MPC problem. Nevertheless, for some of the approaches we formulate adapted versions suitable to address control performance explicitly assuming linear or quadratic stage and terminal cost functions.

4.3.1 Feasibility via enlarging clusters and refinement

The first strategy to compute feasible parametrizations applies the algorithm to compute raw parametrizations (Algorithm 3) to a suitable set of training data and then uses and refines the result. Cluster membership of the training data is used to define sets $\hat{\mathcal{D}}_i$ such that their union contains the desired feasible set $\mathcal{X}_f$, i.e. (4.7) is fulfilled. Then, the quantities M_i, $\mathcal{K}_i$, a_i obtained from the clustering are adjusted to ensure feasibility of $\mathbf{P}_{p_i}(x)$ for x in $\hat{\mathcal{D}}_i$.

Finding sets $\hat{\mathcal{D}}_i$

Assume that Algorithm 3 has been run with training states $\{x_r \in \mathbb{R}^n,\ r = 1, \ldots, s\}$ and given K, q and $\phi(x) = [x^\top, 1]^\top$ and let the resulting output be M_i, $\mathcal{K}_i$, a_i, $i = 1, \ldots, K$ and μ_i^r, $i = 1, \ldots, K$, $r = 1, \ldots, s$. The set $\hat{\mathcal{D}}_i$ is obtained via taking all training states assigned to cluster i, centering a hypercube $\mathcal{C} \subset \mathbb{R}^n$ at each of them and taking the convex hull of the union of all hypercubes and intersecting the result with $\mathcal{X}_f$, i.e. we define

$$\hat{\mathcal{D}}_i = \mathbf{conv}\left(\bigcup_{\{r|\mu_i^r = 1\}} \{x_r\} \oplus \mathcal{C} \right) \cap \mathcal{X}_f, \tag{4.9}$$

where $\oplus$ denotes the Minkowski sum. For a suitable choice of training states and hypercube $\mathcal{C}$, this definition ensures that (4.7) holds.

Lemma 4.4. *Assume the training states $\{x_1, \ldots, x_s\}$ and the hypercube $\mathcal{C}$ are such that*

$$\mathcal{X}_f \subseteq \bigcup_{r=1}^{s} \{x_r\} \oplus \mathcal{C} \tag{4.10}$$

holds and let the sets $\{\hat{\mathcal{D}}_1, \ldots, \hat{\mathcal{D}}_K\}$ be defined via (4.9). Then $\{\hat{\mathcal{D}}_1, \ldots, \hat{\mathcal{D}}_K\}$ is a collection of polytopes for which (4.7) holds with equality, i.e. $\mathcal{X}_f = \cup_{i=1}^{K} \hat{\mathcal{D}}_i$ is satisfied.

Proof. Taking the convex hull of a finite number of polytopes results in a polytope and taking the intersection of two polytopes yields a polytope. Thus, the set $\hat{\mathcal{D}}_i$ is a polytope. The relation "$\subset$" follows from distributive and associative properties of the Minkowski sum and the set union operation whereas the inclusion "$\supset$" follows trivially. $\square$

Generally computing Minkowski sums of polytopes and convex hulls of a set of points is numerically expensive. In (4.9), the Minkowski sums are computed via adding x_r to each one of the 2^n vertices of $\mathcal{C}$ for $\{r|\mu_i^r = 1\}$, then the convex hull is found via identifying the vertices of the resulting set of points and taking the corresponding polytope defined via these vertices. In order to evaluate the intersection, a hyperplane representation of the result is typically needed which is obtained via a numerically expensive facet-enumeration.

Clearly, (4.10) is fulfilled for a suitable choice of training data points and hypercube $\mathcal{C}$. A reasonable choice is to take a regular grid of training data points and to take $\mathcal{C}$ to have twice the distance of two neighboring grid points in each dimension as edge length. This ensures that (4.10) holds in most cases, nevertheless, this choice does not guarantee satisfaction of (4.10) in all cases. Counterexamples can be constructed where the set $\mathcal{X}_f$ has a "spiky" shape at one of its vertices such that it is not properly covered by a regular grid in a vicinity of this vertex. As a remedy, one could reverse the argument and define $\mathcal{X}_f$ via the right hand side of (4.10). Another workaround would be the following: Also grid points could be used which are not contained in $\mathcal{X}_f$ but which are close enough to $\mathcal{X}_f$ so that a hypercube centered at them intersects $\mathcal{X}_f$. For the clustering, these states would have to be represented by a feasible state close to them. Only the intersection of the hypercube centered at the original training point with $\mathcal{X}_f$ would be assigned to the respective cluster.

Similarly, defining $\hat{\mathcal{D}}_i$ via (4.9) is typically compatible with defining $\mathcal{D}_i$ via assigning each state x in $\mathcal{X}_f$ to the same cluster as its closest training state in the sense that $\mathcal{D}_i \subseteq \hat{\mathcal{D}}_i$ holds. Nevertheless, even if a regular grid of training states and a hypercube $\mathcal{C}$ is used which is rather large with respect to the grid of training data, it is possible to construct examples where $\mathcal{D}_i \subset \hat{\mathcal{D}}_i$ does *not* hold. A remedy here would be to define the sets $\mathcal{D}_i$ via assigning each state to the same cluster as the closest training state but only considering clusters for which the corresponding set $\hat{\mathcal{D}}_i$ contains the state.

Remark 4.5. *Intersecting in (4.9) the convex hull with $\mathcal{X}_f$ is done merely to avoid having sets $\hat{\mathcal{D}}_i$ which are unnecessarily large as this would render achieving feasibility of $\boldsymbol{P}_{p_i}(x)$ for all x in $\hat{\mathcal{D}}_i$ (in a next step below) unnecessarily hard. Generally, the intersection could be omitted as long as $\hat{\mathcal{D}}_i \subseteq \mathbb{X}_N$ is ensured as this is a necessary condition for the existence of M_i, $\mathcal{K}_i$, a_i such that $\boldsymbol{P}_{p_i}(x)$ is feasible for all x in $\hat{\mathcal{D}}_i$. Omitting the intersection simplifies evaluation of (4.9) and in numerical simulations generally also simplified the shape of the resulting sets $\hat{\mathcal{D}}_i$ in the sense that they are defined by less vertices and less hyperplanes, respectively.*

Next, we address refinement of M_i, $\mathcal{K}_i$ and a_i to ensure feasibility of $\boldsymbol{P}_{p_i}(x)$ for x in $\hat{\mathcal{D}}_i$.

Refinement of M_i, $\mathcal{K}_i$, a_i

Refining the quantities M_i, $\mathcal{K}_i$ and a_i follows the strategy to render the corresponding parametrized optimization problems $\boldsymbol{P}_{p_i}(x)$ feasible at the vertices of the polytope $\hat{\mathcal{D}}_i$ which due to convexity of the constraints implies feasibility for all x in $\hat{\mathcal{D}}_i$. We first present a version of the refinement which addresses feasibility only. Let $\{x_1, \ldots, x_{n_R} \in \mathcal{X}_f\}$ and consider for given $\{\tilde{U}_1, \ldots, \tilde{U}_{n_R} \in \mathbb{R}^q\}$ the linear program

$$\min_{M,\mathcal{K},a,\beta} \beta$$
$$\text{s.t. } GM\tilde{U}_r \le (E - G\mathcal{K})x_r + \beta W + Ga, \quad r = 1, \ldots, n_R, \quad \beta \ge 1 \tag{4.11}$$

where $M \in \mathbb{R}^{mN \times q}$, $\mathcal{K} \in \mathbb{R}^{mN \times n}$ and $a \in \mathbb{R}^{mN}$. The following holds.

Lemma 4.6 (Feasibility of $\boldsymbol{P}_{p_i}(x)$). *Consider (4.11) and let the vertices of $\hat{\mathcal{D}}_i$ be contained in $\{x_1, \ldots, x_{n_R}\}$. If (4.11) is feasible with M^*, $\mathcal{K}^*$, a^* and $\beta^* = 1$ then using $M_i = M^*$, $\mathcal{K}_i = \mathcal{K}^*$ and $a_i = a^*$ in part p_i of the parametrization ensures feasibility of $\boldsymbol{P}_{p_i}(x)$ for all x in $\hat{\mathcal{D}}_i$.*

Proof. The origin is in the respective interior of the state, input and terminal constraint set and ($\bar{x} = 0$, $\bar{u} = 0$) is a steady state of the system. Thus, $U = 0$ is a solution of (4.5) for $x = 0$ at which no constraints are active. Hence, all elements in vector W have to be strictly positive. Therefore, feasibility of (4.11) with $\beta \leq 1$, implies that $G(M\tilde{U}_r + \mathcal{K}x_r + a) \leq W + Ex_r$ holds. In particular, $\tilde{U}_r$ is a feasible solution of $\mathbf{P}_{p_i}(x_r)$. This implies feasibility of $\mathbf{P}_{p_i}(\cdot)$ at the vertices of $\hat{\mathcal{D}}_i$. Feasibility for all x in $\hat{\mathcal{D}}_i$ follows from convexity of the constraints of $\mathbf{P}_{p_i}(\cdot)$. □

As the result of this optimization strongly depends on the values for $\{\tilde{U}_1, \ldots, \tilde{U}_{n_R}\}$ used therein, iteratively updating these values and updating M, $\mathcal{K}$ and a via (4.11) is reasonable. The $\tilde{U}$ update can be done via solving for each element in $\{x_1, \ldots, x_{n_R}\}$ the linear program

$$\min_{\tilde{U}, \beta} \beta$$
$$\text{s.t.} \quad GM\tilde{U} \leq (E - G\mathcal{K})x_r + \beta W + Ga, \quad \beta \geq 1 \tag{4.12}$$

where the current values of M, $\mathcal{K}$ and a are fixed. Both of the latter optimization problems are linear programs and hence can be solved very efficiently.

Investing slightly higher numerical effort and assuming a linear or quadratic objective function in the underlying mpP, the refining optimization can be used to directly address open-loop performance for some sampled states. In this case instead of (4.11) the optimization

$$\min_{M, \mathcal{K}, a, \beta} \gamma\beta + \sum_{r=1}^{n_R} J(x_r, M\tilde{U}_r + \mathcal{K}x_r + a)/V^*(x_r)$$
$$\text{s.t.} \quad GM\tilde{U}_r \leq (E - G\mathcal{K})x_r + \beta W + Ga, \quad r = 1, \ldots, n_R, \quad \beta \geq 1 \tag{4.13}$$

is solved. The $\tilde{U}$ update in this case is done via solving

$$\min_{\tilde{U}, \beta} \gamma\beta + J(x_r, M\tilde{U}_r + \mathcal{K}x_r + a)$$
$$\text{s.t.} \quad GM\tilde{U} \leq (E - G\mathcal{K})x_r + \beta W + Ga, \quad \beta \geq 1. \tag{4.14}$$

In both optimization problems, the weight γ on minimizing β is chosen as a large constant to achieve feasibility with high priority; the constraint $\beta \geq 1$ ensures that once feasibility has been achieved, open-loop performance is improved. Thus, the latter optimization problems are basically soft constrained versions of a performance optimization. Among different ways to implement the constraint softening, the proposed realization thereof was chosen as it performed best in numerical tests. As (4.13) comprises all constraints of (4.11), Lemma 4.6 holds correspondingly with respect to (4.13). If the underlying mpP has a linear objective function, the latter two optimization problems are linear programs, if the mpP has a convex quadratic objective function, the latter two optimization problems are convex quadratic programs.

Remark 4.7. *Note that if only $\beta^* > 1$ is achieved in the refining optimization, using $p_i(x, \tilde{U}) = M^*\tilde{U} + \mathcal{K}^*x + \frac{1}{\beta^*}a^*$, feasibility on the scaled set $\frac{1}{\beta^*}\hat{\mathcal{D}}_i = \{\frac{1}{\beta^*}x | x \in \hat{\mathcal{D}}_i\}$ is achieved. If for $\beta_0 > 0$ it holds for all parts i of the parametrization that $\beta^* \leq \beta_0$, this implies that $\frac{1}{\beta_0}\mathcal{X}_f$ is contained in the union of the scaled sets $\frac{1}{\beta^*}\hat{\mathcal{D}}_i$ and overall a parametrization which is feasible on $\frac{1}{\beta_0}\mathcal{X}_f$ is obtained.*

The iterative refinement procedure for the quantities M_i, $\mathcal{K}_i$, a_i is summarized in Algorithm 6.

Algorithm 6 Refinement of M_i, $\mathcal{K}_i$, a_i

Input: States $\{x_r \in \hat{\mathcal{D}}_i,\ r = 1, \ldots, n_S\}$ which comprise vertices of $\hat{\mathcal{D}}_i$, maximum iterations
    $i_{\max}$, initial values M_i^0, $\mathcal{K}_i^0$, a_i^0

1: set $it = 0$
2: update $\tilde{U}_r$, $r = 1, \ldots, n_R$ via (4.12) (or (4.14)) using M_i^0, $\mathcal{K}_i^0$, a_i^0
3: update M_i, $\mathcal{K}_i$, a_i, β via (4.11) (or (4.13))
4: **while** $\beta > 1$ and $it < i_{\max}$ **do**
5:    set $it = it + 1$
6:    update $\tilde{U}_r$, $r = 1, \ldots, n_R$ via (4.12) (or (4.14))
7:    update M_i, $\mathcal{K}_i$, a_i, β via (4.11) (or (4.13))
8: **end while**
9: **if** $\beta > 1$ **then**
10:    terminate // refinement infeasible
11: **end if**
Output: If refinement feasible: M_i, $\mathcal{K}_i$, a_i

Overall procedure

Summarizing, in the first approach to compute a feasible parametrization, first the tailored
subspace clustering algorithm is run, second the sets $\hat{\mathcal{D}}_i$ are computed via (4.9) based on
the output of the tailored subspace clustering algorithm and finally M_i, $\mathcal{K}_i$ and a_i are
refined via either iterating over (4.11) and (4.12) or iterating over (4.13) and (4.14) until
feasibility is achieved ($\beta \leq 1$) or a maximum number of iterations is reached. Whereas
feasibility of the refinement is not guaranteed, it can typically be achieved for sufficiently
large values of the hyperparameters K and q. The complete procedure is formalized in
Algorithm 7.

Algorithm 7 Feasible parametrization via enlarging sets and refinement

---

Input: training states $\{x_r \in \mathbb{R}^n,\ r = 1, \ldots, s\}$, mpP (4.2), K, q, $\mathcal{X}_f$, hypercube $\mathcal{C}$ such
    that (4.10) holds, maximum iterations $i_{\max}$

1: compute $U_r = U^*(x_r)$, $r = 1, \ldots, s$ via solving (4.2)
2: run Algorithm 2 with $y_r = \phi(x_r) = [x_r^\top, 1]^\top$, K, q. obtain M_i^0, $\mathcal{K}_i^0$, a_i^0, $i = 1, \ldots, K$
    and μ_i^r, $i = 1, \ldots, K$, $r = 1, \ldots, s$
3: **for** $i = 1, \ldots, K$ **do**
4:    compute $\hat{\mathcal{D}}_i$ via (4.9)
5:    run Algorithm 6 using $\{x_r \mid \mu_i^r = 1\} \cup \text{vert}(\hat{\mathcal{D}}_i)$, and M_i^0, $\mathcal{K}_i^0$, a_i^0, $i_{\max}$ as input.
6:    **if** Algorithm 6 terminated successfully **then**
7:        obtain M_i, $\mathcal{K}_i$, a_i
8:    **else**
9:        terminate // refinement infeasible
10:    **end if**
11: **end for**
Output: If refinement feasible: $\hat{\mathcal{D}}_i$, M_i, $\mathcal{K}_i$, a_i, $i = 1, \ldots, K$

Proposition 4.8. *Algorithm 7, yields a parametrization which is feasible on $\mathcal{X}_f$ if it
terminates successfully.*

Proof. This follows from Lemma 4.4 and Lemma 4.6. □

4.3.2 Feasibility via clustering facets and refinement

In the first approach for finding a feasible parametrization, the desired feasible set $\mathcal{X}_f$ was implicitly included into the clustering via clustering training states sampled from the desired feasible set and generalizing cluster membership to sufficiently large hypercubes around the training states. For the approach to be presented next, clustering of the desired feasible set is included into the clustering procedure more explicitly. This is achieved by including the facets of the desired feasible set $\mathcal{X}_f$ into the clustering procedure. The sets $\hat{\mathcal{D}}_i$ can in this case be taken as the convex hull of the states assigned to the cluster, the origin and possibly facets which have been assigned to the respective cluster. Following this approach, it is much simpler to determine the sets $\hat{\mathcal{D}}_i$ and the procedure scales better in the state dimension than it is the case for the first approach. To be more precise, in this approach not complete facets are clustered but they are subdivided into simplices prior to the clustering and the resulting simplices are clustered. This increases flexibility in the shape of the sets $\hat{\mathcal{D}}_i$ and enables to establish a feasibility guarantee in the sequel. For the clustering, the simplices are represented via their vertices together with the constraint that all vertices of one simplex have to be assigned to the same cluster. To this end, we first extend the tailored subspace clustering algorithm introduced above by the possibility to enforce constraints on cluster membership of the training data. In a second step, we then introduce the complete procedure to find a feasible parametrization based on clustering facets of the desired feasible set.

Tailored subspace clustering algorithm including constraints on cluster membership

The subspace clustering algorithm to be presented next is directly derived from the general tailored subspace clustering algorithm introduced above in Algorithm 2, Section 3.3.2. In the following we mainly focus on differences between the algorithms to keep the presentation concise. As motivated above, the additional feature required for the subspace clustering algorithm is the possibility to enforce constraints with respect to cluster membership of the clustered data points. Let the data point pairs to be clustered be given by $\{(y_r, U_r),\ r = 1, \ldots, s\}$, then the constraints regarding cluster membership are given in the form that all pairs with index r in the same index set $I_t \subset \{1, \ldots, s\}$ are to be assigned to the same cluster. Different index sets $\{I_1, \ldots, I_{n_c}\}$ are used and not each index has to be contained in one of the index sets, i.e., some data point pairs can be assigned freely. The optimization problem which is addressed by the constrained subspace clustering algorithm is given by

$$\min_{M_i, N_i, \tilde{U}_r, \mu_i^r,\ i=1,\ldots,K,\ r=1,\ldots,s} J_C \tag{4.15a}$$

$$\text{s.t. } \mu_i^r \in \{0, 1\} \text{ and } \sum_{i=1}^{K} \mu_i^r = 1, \tag{4.15b}$$

$$\mu_i^r = \overline{\mu}_i^t,\ r \in I_t,\ i = 1, \ldots, K,\ t = 1, \ldots, n_S \tag{4.15c}$$

with

$$J_C = \sum_{i=1}^{K} \sum_{r=1}^{s} \mu_i^r \| U_r - M_i \tilde{U}_r - N_i y_r \|_2$$

and where $M_i \in \mathbb{R}^{mN \times q}$, $N_i \in \mathbb{R}^{mN \times b}$ and $\tilde{U}_r \in \mathbb{R}^q$. Therein, (4.15c) expresses the constraint on cluster membership of the data points.

The strategy we suggest to find a good approximation of the solution to this problem is again to iteratively execute a cluster update step and a cluster assignment step. In this case, the cluster assignment step solves for fixed cluster specific matrices M_i, N_i the optimization problem

$$\min_{\mu_i^r, \tilde{U}_r} \sum_{i=1}^{K} \sum_{r=1}^{s} \mu_i^r \|U_r - M_i \tilde{U}_r - N_i y_r\|_2 \tag{4.16}$$

s.t. (4.15b) and (4.15c).

This optimization can be solved efficiently via an exhaustive enumeration strategy. First $d_{i,r} = \|U_r - N_i y_r - M_i M_i^+ (U_r - N_i y_r)\|_2^2$ is computed for all i, r. Then, $\bar{d}_{i,t} = \sum_{r \in I_t} d_{i,r}$ is evaluated. Finally, $\mu_i^r = 1$ is set either for $i = \arg\min_i \bar{d}_{i,t}$ if $r \in I_t$ or based on $i = \arg\min_i d_{i,r}$. The problem solved in the cluster update step is identical to the one in the unconstrained case and is given by

$$\min_{M_i, N_i, \tilde{U}_r} \sum_{\{r | \mu_i^r = 1\}} \|U_r - M_i \tilde{U}_r - N_i y_r\|_2. \tag{4.17}$$

The resulting procedure is completely analogous to the non-constrained case and is stated in Algorithm 8. We will below require the index sets to be pairwise disjoint in order to

---

Algorithm 8 Constrained tailored subspace clustering algorithm

---

Input: pairs $\{(y_r, U_r) \in \mathbb{R}^b \times \mathbb{R}^{mN}, r = 1, \ldots, s\}$, $\{I_t, t = 1, \ldots, n_S\}$, parameters $K, q \in \mathbb{N}$
  1: initialize μ_i^r, $i = 1, \ldots, K$, $r = 1, \ldots, s$ randomly subject to (4.15b) and (4.15c)
  2: update matrices M_i, N_i via (4.17), for $i = 1, \ldots, K$
  3: update μ_i^r, $i = 1, \ldots, K$, $r = 1, \ldots, s$ via (4.16)
  4: if μ_i^r has changed w.r.t. previous iteration, go back to line 2. Else: done
Output: matrices M_i, N_i, $i = 1, \ldots, K$, cluster assignment μ_i^r, $i = 1, \ldots, K$, $r = 1, \ldots, s$

---

ensure existence of meaningful solutions. Exactly the same convergence results as in the non-constrained case can be established for this algorithm, applying identical arguments. See Proposition 3.5 in Section 3.3.2 and the discussion thereafter.

Training data and definition of sets $\hat{\mathcal{D}}_i$

The training data used in this approach to compute feasible parametrizations comprises two types of states: Firstly, states representing simplices at the facets of $\mathcal{X}_f$ (together with the corresponding index sets) and, secondly, additional states sampled from the interior of $\mathcal{X}_f$. Inclusion of the desired feasible set in the union of the sets $\hat{\mathcal{D}}_i$ is ensured solely by the first type of states whereas the second type of states contributes to approximation accuracy of solutions in the interior of $\mathcal{X}_f$ and, thus, eventually to control performance. Hence, they can be selected totally independently of the desired feasible set and we only address the procedure to find the first type of states together with the respective index sets more in detail in the following.

Given a desired feasible set $\mathcal{X}_f$, first its facets $\{\mathcal{F}_j, j = 1, \ldots, n_{\mathcal{F}}\}$ are required in vertex representation. As $\mathcal{X}_f$ is a full-dimensional polytope in $\mathbb{R}^n$, its facets are $(n-1)$-dimensional polytopes. Applying a Delaunay[1] triangulation to the vertices of the facet, each facet $\mathcal{F}_j$ is then decomposed into a set of $(n-1)$-dimensional simplices $\{\mathcal{S}_t,\ t=1,\ldots,n_{\mathcal{S}j}\}$. Overall the boundary $\partial \mathcal{X}_f$ of $\mathcal{X}_f$ is recovered as

$$\partial \mathcal{X}_f = \bigcup_{j=1}^{n_{\mathcal{F}}} \mathcal{F}_j = \bigcup_{t=1}^{n_{\mathcal{S}}} \mathcal{S}_t = \bigcup_{t=1}^{n_{\mathcal{S}}} \mathrm{conv}\left(\{x_t^1,\ldots,x_t^n\}\right) \qquad (4.18)$$

where $\{x_t^1,\ldots,x_t^n\}$ are the vertices of simplex $\mathcal{S}_t$ and $n_{\mathcal{S}}$ is the total number of simplices. The set of training states $\mathcal{T}$ is then taken as the union of all simplex vertices plus additional training states sampled from $\mathcal{X}_f$

$$\mathcal{T} = \{x_t^i, i=1,\ldots,n,\ t=1,\ldots,n_{\mathcal{S}}\} \cup \{x_r^{\mathrm{int}} \in \mathcal{X}_f | r = 1,\ldots,n_a\}. \qquad (4.19)$$

Furthermore, for each simplex $\mathcal{S}_t$ one index set I_t is defined such that its vertices are recovered as

$$\{x_r \in \mathcal{T} | r \in I_t\} = \{x_t^1,\ldots,x_t^n\} \qquad (4.20)$$

and such that the index sets are pairwise disjoint, $I_t \cap I_{\tilde{t}} = \varnothing$ if $t \neq \tilde{t}$. This requires the set $\mathcal{T}$ to contain as many instances of the same element as this element plays the role of a vertex of one of the simplices. It remains to compute the optimal predicted input sequence for each element in $\mathcal{T}$ and apply the constrained clustering algorithm to the data. In this case, the sets $\hat{\mathcal{D}}_i$ can be defined directly based on the clustering result according to

$$\hat{\mathcal{D}}_i = \mathrm{conv}\left(\bigcup_{\{r|\ \mu_i^r=1\}} \{x_r\} \cup \{0\} \right). \qquad (4.21)$$

In particular, here only the convex hull of a set of points is to be taken to obtain the sets $\hat{\mathcal{D}}_i$ and the following holds.

Lemma 4.9. *Consider a set of training states $\mathcal{T}$ such that (4.18) and (4.19) hold and let a corresponding cluster assignment μ_i^r, $r=1,\ldots,s$, $i=1,\ldots,K$ be given such that for all simplices $\mathcal{S}_t$ there is a cluster i for which*

$$\{x_t^1,\ldots,x_t^n\} \subset \{x_r | \mu_i^r = 1\} \qquad (4.22)$$

holds. Then $\hat{\mathcal{D}}_i$ defined via (4.21) is a polytope such that overall (4.7) holds with equality, i.e., $\mathcal{X}_f = \cup_{i=1}^K \hat{\mathcal{D}}_i$ is satisfied.

Note that (4.22) simply means that all vertices of a simplex are assigned to the same cluster. This constraint is ensured by constraint (4.15c).

Proof. Let x be any point in $\mathcal{X}_f$. There exists $\alpha \geq 1$ such that $\alpha x \in \partial \mathcal{X}_f$ as $\mathcal{X}_f$ is a bounded polytope with zero in its interior. According to (4.18) this implies that $\alpha x \in \mathcal{S}_t$ for one simplex $\mathcal{S}_t$. This simplex is contained in one of the clusters as (4.22) holds and due to (4.21) it is contained in the corresponding set $\hat{\mathcal{D}}_i$. Due to convexity also any convex combination of αx and 0 is contained in $\hat{\mathcal{D}}_i$, in particular $x \in \hat{\mathcal{D}}_i$. It results that $\mathcal{X}_f \subseteq \cup_{i=1}^K \hat{\mathcal{D}}_i$. On the other hand, the sets $\hat{\mathcal{D}}_i$ are defined as convex hulls of points sampled from $\mathcal{X}_f$. As $\mathcal{X}_f$ is convex, this implies $\mathcal{X}_f \supset \cup_{i=1}^K \hat{\mathcal{D}}_i$. $\qquad \square$

---

[1] Whereas clearly any triangulation could be used, the Delaunay triangulation yields a result which is optimal in some respects, for example, fostering triangles of regular shape. Here, having simplices of regular shape can generally reduce conservatism.

Overall procedure

In this approach to compute feasible parametrizations, first suitable training data has to be generated which represents a triangulation of the facets of the desired feasible set in form of vertices together with suitable index sets grouping the vertices into n-tuples corresponding to the simplices. Additional training data sampled from the interior of the desired feasible set can be added. The constrained subspace clustering algorithm is then applied and it guarantees that all vertices of one specific simplex are assigned to the same cluster. This ensures that taking the sets $\hat{\mathcal{D}}_i$ as the convex hull of the origin and all states in the corresponding cluster, the union of the resulting sets $\hat{\mathcal{D}}_i$ automatically equals the desired feasible set $\mathcal{X}_f$. Finally, M_i and $\mathcal{K}_i$ have to be refined to guarantee feasibility on the sets $\hat{\mathcal{D}}_i$ applying the same procedure as before (Algorithm 6). The whole procedure is summarized in Algorithm 9.

Algorithm 9 Feasible parametrization via clustering facets and refinement

---

Input: mpP (4.2), K, q, $\mathcal{X}_f$, additional training data $\{x_1^{\text{int}}, \ldots, x_{n^a}^{\text{int}}\}$, maximum iterations $i_{\max}$
1: find training states $\{x_1, \ldots, x_s\}$ according to (4.18) and (4.19)
2: define index sets such that (4.20) holds
3: compute $\{U_r = U^*(x_r),\ r = 1, \ldots, s\}$ via solving (4.2)
4: run Algorithm 8 with $y_r = \phi(x_r) = [x_r^\top, 1]^\top$, K, q, obtain M_i^0, $\mathcal{K}_i^0$, a_i^0, $i = 1, \ldots, K$ and μ_i^r, $i = 1, \ldots, K$, $r = 1, \ldots, s$
5: **for** $i = 1, \ldots, K$ **do**
6:   compute $\hat{\mathcal{D}}_i$ via (4.21)
7:   run Algorithm 6 using $\{x_r|\ \mu_i^r = 1\} \cup \{0\}$ and M_i^0, $\mathcal{K}_i^0$, a_i^0, $i_{\max}$ as input.
8:   **if** Algorithm 6 terminated successfully **then**
9:     obtain M_i, $\mathcal{K}_i$, a_i
10: **else**
11:    terminate // refinement infeasible
12: **end if**
13: **end for**
Output: If refinement feasible: $\hat{\mathcal{D}}_i$, M_i, $\mathcal{K}_i$, a_i, $i = 1, \ldots, K$

Proposition 4.10. *Algorithm 9 yields a parametrization which is feasible on $\mathcal{X}_f$ if it terminates successfully.*

Proof. Relations (4.18) – (4.20) together with constraint (4.15c) ensure that (4.22) holds. The training data used satisfies (4.18) – (4.20) and during all iterations of Algorithm 8 constraint (4.15c) is satisfied. Thus, Lemma 4.9 and Lemma 4.6 apply and yield the claim. $\square$

Again feasibility of the refinement cannot be guaranteed generally and the proposed method is rather to be applied in an iterative fashion to find suitable hyperparameters K and q. Nevertheless, feasibility guarantees for two limiting extreme cases of values for K and q can be given.

Proposition 4.11. *Consider Algorithm 9 and the refinement executed in line 7. Refining Algorithm 6 is feasible for ($K = 1$, $q = mN$) using $\tilde{U}_r = U_r$ and it is feasible for ($K = n_{\mathcal{S}}$, $q = 0$) assuming $\hat{\mathcal{D}}_i \subseteq conv(\{x_r | r \in I_i\} \cup \{0\})$ for $i = 1, \dots, n_{\mathcal{S}}$.*

Proof. The first part is obvious from setting $M_i = I$ and $\mathcal{K}_i = 0$. In the second case, a feasible solution is given by $\mathcal{K}_i = [U(x_i^1) \dots U(x_i^n)][x_i^1 \dots x_i^n]^{-1}$, $a_i = 0$, where $U(x_i^j)$ is any feasible input sequence for x_i^j. This $\mathcal{K}_i$ yields for each x in $conv(\{x_r | r \in I_i\} \cup \{0\})$ a convex combination of the input sequences at the vertices which is feasible by convexity. Existence of the inverse follows from $\{0, x_i^1, \dots, x_i^n\}$ being the vertices of a simplex with volume greater than zero. $\square$

The argument for the case ($K = n_{\mathcal{S}}$, $q = 0$) founds on ideas which have been used for interpolation based schemes, see e.g. (Valencia-Palomo and Rossiter, 2011) and basically recovers an interpolated solution. In fact the numbers given in the theorem are of theoretic nature. In all considered examples considerably simpler feasible parametrizations could be found, see also the examples presented below in this thesis.

Remark 4.12. *Including the origin in the definition of $\hat{\mathcal{D}}_i$ (4.21) might seem arbitrary to some extent. In fact there is a good reason for that. In practice, parametrizations for linear systems for which the constant part has been set to zero ($a_i = 0$) have turned out to yield best results. When using such parametrization, the parametrized optimization is always feasible at $x = 0$ as $U = 0$ is in the image of the parametrization for $x = 0$. As a result, enforcing feasibility at $x = 0$ does in fact not introduce any additional requirement nor conservatism.*

Exploiting symmetry

Many linear MPC problems possess symmetric constraints which result in feasible sets of the corresponding mpP being point symmetric with respect to the origin. If in such case a parametrization for linear systems without constant term ($a_i = 0$) is applied, the feasible set remains symmetric:

Lemma 4.13. *Consider an mpP derived from an MPC problem for which the state, terminal and input constraints are point symmetric with respect to the origin, i.e., $x \in \mathbb{X} \Leftrightarrow -x \in \mathbb{X}$, $x \in \mathbb{X}_T \Leftrightarrow -x \in \mathbb{X}_T$ and $u \in \mathbb{U} \Leftrightarrow -u \in \mathbb{U}$. Let $\boldsymbol{P}_{p_i}$ be the corresponding parametrized mpP using a parametrization for linear systems for which $a_i = 0$ holds. The set of states for which $\boldsymbol{P}_{p_i}(x)$ is feasible is point symmetric with respect to the origin, i.e, $\boldsymbol{P}_{p_i}(x)$ is feasible if and only if $\boldsymbol{P}_{p_i}(-x)$ is feasible.*

Proof. Let $\tilde{U}$ be a feasible solution of $\boldsymbol{P}_{p_i}(x)$. Thus, $U = p_i(x, \tilde{U})$ is a stacked vector of inputs each satisfying the input constraints and it results via $X = \hat{A}x + \hat{B}U$ (for suitable matrices $\hat{A} \in \mathbb{R}^{nN \times n}$ and $\hat{B} \in \mathbb{R}^{nN \times mN}$) in a stacked vector of states of which each one is compatible with the state constraints and the last element satisfies the terminal constraints. Correspondingly, $-\tilde{U}$ results for $-x$ in $-U = p_i(-x, -\tilde{U})$, which is a stacked vector of inputs satisfying each the input constraints and which yields the corresponding stacked vector of states $-X = -\hat{A}x - \hat{B}U$ of which, by assumption, each element is compatible with the state constraints and the last element satisfies the terminal constraints. In particular, $-\tilde{U}$ is a feasible solution of $\boldsymbol{P}_{p_i}(-x)$. $\square$

Figure 4.1: Two equivalent Delaunay triangulations of a square.

On the one hand, for symmetric problems and parametrizations for which $a_i = 0$ holds this fact can be exploited to reduce the size of the refining optimization: If the parametrization is rendered feasible for state x, feasibility for state $-x$ is automatically obtained and state $-x$ can be neglected in the refinement procedure (and removed in cases where it would be included otherwise). On the other hand, feasible sets of the parametrized optimization are inherently shaped point symmetrically with respect to the origin. In return, in order to reduce conservatism, it is important to find sets $\hat{\mathcal{D}}_i$ which are point symmetric with respect to the origin. Rendering a non-symmetric set $\hat{\mathcal{D}}_i$ feasible would in fact mean to render $\hat{\mathcal{D}}_i$ feasible and at the same time a second set which is located point symmetrically to $\hat{\mathcal{D}}_i$ with respect to the origin. Overall this would mean to have unnecessarily overlapping sets $\hat{\mathcal{D}}_i$ which are all rendered feasible making the refinement procedure more challenging than necessary. A problem of the proposed method in this respect is that it does not automatically admit such symmetric sets. This is due to the fact that the Delaunay triangulation is not unique. Consider Figure 4.1 which shows two equivalent Delaunay triangulations of a square. Imagine the squares to be opposite facets of a cube $\mathcal{X}_f$. Then, the set $\hat{\mathcal{D}}_i$ determined via (4.21) which contains any combination of triangles located at opposite facets could never be shaped point symmetrically with respect to the origin and $\hat{\mathcal{D}}_i$ would be largely overlapping with another set $\hat{\mathcal{D}}_j$.

In fact a numerical procedure to compute the Delaunay triangulation of the facets provided in MATLAB turned out to yield largely non-symmetric solutions when applied to a general polytope $\mathcal{X}_f$. Thus, a post processing step in order to achieve symmetry was needed. Surprisingly, first of all for slightly higher dimensional problems, this turned out to be challenging for numerical reasons. A solution approach was developed during the work on this thesis. It essentially clusters the surface normals of the simplices in order to organize them in a first step and then divides them into symmetric simplices.

4.3.3 Feasibility via independent part

Let us now come to the third approach to compute feasible parametrizations for linear systems. The second approach to find feasible parametrizations was simpler than the first one and scales better in the state dimension than the first one. Yet, in the algorithm still some operations are required which become numerically expensive with growing state dimension and many training states are required in the clustering procedure if the desired feasible set $\mathcal{X}_f$ comprises many vertices.

In order to further alleviate these drawbacks, we next introduce an approach to obtain a feasible parametrization which separates the offline computations into a first part which ensures feasibility and a second part which contributes control performance. The idea for the feasibility implying first part is as follows. Find in a systematic yet simple way a linear mapping $x \mapsto \mathcal{K}x$, with $\mathcal{K} \in \mathbb{R}^{mN \times n}$, such that there is a large and possibly pre-specified set of states $\mathcal{X}_f \subset \mathbb{R}^n$ such that for each $x \in \mathcal{X}_f$ the constraints of (4.2) are satisfied by

$U = \mathcal{K}x$. As a consequence, any parametrization of the form $p_i(x, \tilde{U}) = M_i\tilde{U} + \mathcal{K}x$ is feasible at least for all $x \in \mathcal{X}_f$ simply by using $\tilde{U} = 0$. The matrices M_i are then computed in a second step solely to improve approximation accuracy of optimal input sequences. We call $\mathcal{K}x$ the feasibility part and $M_i\tilde{U}$ the performance part. Following this approach, a constant offset is not included into the parametrization.

The feasibility part can be determined via the linear program

$$\min_{\mathcal{K},\beta_1,\ldots,\beta_s} \sum_{r=1}^{s} \beta_r \tag{4.23}$$
$$\text{s.t. } G\mathcal{K}x_r \leq Ex_r + \beta_r W, \ \beta_r \geq 0, \ r = 1,\ldots,s,$$

where $\mathcal{K} \in \mathbb{R}^{mN \times n}$, $\{\beta_1,\ldots,\beta_s \in \mathbb{R}\}$ holds and $\{x_1,\ldots,x_s \in \mathbb{R}^n\}$ are given states. Some statements about the resulting feasible set can be made.

Proposition 4.14. *Let $\mathcal{K}^*$ and $\{\beta_1^*,\ldots,\beta_s^*\}$ be minimizer of (4.23). The following holds.*

- *The set of states x for which $U = \mathcal{K}^*x$ is an admissible input sequence satisfying the constraints of (4.2), is given by*

$$\hat{\mathcal{X}}_f = \{x|(G\mathcal{K}^* - E)x \leq W\}. \tag{4.24}$$

- *It holds that*

$$conv\left(\left\{\frac{1}{\beta_r^*}x_r|\beta_r^* > 0\right\}\right) \subseteq \hat{\mathcal{X}}_f \tag{4.25}$$

and the feasible set $\hat{\mathcal{X}}_f$ is unbounded in the directions $\{x_r|\beta_r^* = 0\}$.

Proof. Equation (4.24) follows directly from plugging in $U = \mathcal{K}^*x$ into the constraints of (4.2). Consider r with $\beta_r^* > 0$. Relation (4.25) follows from dividing the constraints of (4.23) row-wise by β_r^* and by convexity of the constraints. Second, consider r with $\beta_r^* = 0$. This implies $(G\mathcal{K}^* - E)x_r \leq 0$ and $(G\mathcal{K}^* - E)(\alpha x_r) \leq 0$ for all $\alpha \geq 0$. This shows unboundedness of $\hat{\mathcal{X}}_f$ in direction x_r. $\quad\square$

Note that unbounded feasible sets are in fact possible. Example 3.19 above shows such situation as no state constraints and no terminal constraints are imposed. Hence, $\mathcal{K}^* = 0$ and $\beta_1^* = \ldots = \beta_s^* = 0$ minimizes (4.23) in this case. The next result follows trivially.

Corollary 4.15. *Consider a parametrization for linear systems where $p_i(x, \tilde{U}) = M_i\tilde{U} + \mathcal{K}^*x$ holds for arbitrary M_i and where $\mathcal{K}^*$ is an optimizer of (4.23). It holds that*

$$\hat{\mathcal{X}}_f \subseteq \left\{x|\boldsymbol{P}_{p_i}(x) \text{ is feasible}\right\}. \tag{4.26}$$

with $\hat{\mathcal{X}}_f$ defined in (4.24).

In particular, the feasible set of $\boldsymbol{P}_{p_i}$ is available in hyperplane representation, given in (4.24), and an inner approximation thereof is available in vertex representation, given in (4.25). This can be very convenient and largely simplify matters, first of all for problems with larger state dimension.

Extensions of (4.23) to weighting the single states $\{x_1, \ldots, x_s\}$ differently via weighting the β_r variables in the objective function, to additionally minimizing a common upper bound on the β_r variables or to using only one common β for all states are straight forward. In the latter case, it holds that if a linear mapping $U = \mathcal{K}x$ exists such that conv $(\{x_1, \ldots, x_s\}) \subseteq \hat{\mathcal{X}}_f$, this is achieved by any solution of (4.23).

The complete procedure to compute parametrizations via this approach is then as follows. First, find states $\{x_1, \ldots, x_s\}$ which represent the desired feasible set $\mathcal{X}_f$ suitably. A typical choice would be to take the vertices of $\mathcal{X}_f$. Then the linear program (4.23) is solved for these states such that $\mathcal{K}^*$ is obtained. In the parametrization, $\mathcal{K}_i = \mathcal{K}^*$, $i = 1, \ldots, K$ is used. Only the matrices M_i are computed via the tailored subspace clustering procedure, where $\phi \equiv 0$ is set. To compensate the linear part $\mathcal{K}_i x$ which is in fact present in the parametrization, transformed training data is used for clustering: Instead of $U^*(x_r)$ here $U^*(x_r) - \mathcal{K}^* x_r$ is used. Finally, for the parametrization the matrices M_i are taken from the clustering. The sets $\hat{\mathcal{D}}_i = \hat{\mathcal{X}}_f$ are defined with $\hat{\mathcal{X}}_f$ according to (4.24). The procedure is summarized in Algorithm 10 and feasibility of the resulting parametrization is characterized in Corollary 4.15.

Algorithm 10 Feasible parametrization via independent part and clustering

---

Input: training states $\{x_1, \ldots, x_s \in \mathbb{R}^n\}$, mpP (4.2), K, q

1: solve LP (4.23) using $\{x_1, \ldots, x_s\}$ and the constraints from mpP (4.2), obtain optimizer $\mathcal{K}^*$, $\beta_1^*, \ldots, \beta_s^*$
2: compute $U_r = U^*(x_r) - \mathcal{K}^* x_r$, $r = 1, \ldots, s$ where $U^*(x_r)$ solves (4.2) for x_r
3: run Algorithm 2 with $y_r = 0$, K, q. obtain M_i, $i = 1, \ldots, K$ and μ_i^r, $i = 1, \ldots, K$, $r = 1, \ldots, s$
4: define $\mathcal{K}_i = \mathcal{K}^*$, $i = 1, \ldots, K$
5: define $\hat{\mathcal{D}}_i = \hat{\mathcal{X}}_f$, $i = 1, \ldots, K$ with $\hat{\mathcal{X}}_f$ defined in (4.25)

Output: sets $\hat{\mathcal{D}}_i$, matrices M_i, $\mathcal{K}_i$, $i = 1, \ldots, K$

In this approach, the most important tuning knob to find a parametrization which is feasible for a given desired feasible set $\mathcal{X}_f$ is the prediction horizon N used. Under reasonable assumptions, increasing the prediction horizon results in a non-increasing sequence of objective function values in linear program (4.23) and thus generally in feasible sets of increasing size.

Lemma 4.16. *Consider linear program (4.23) and assume that in the underlying MPC problem the terminal control law has the form $\kappa(x) = K_T x$ for matrix $K_T \in \mathbb{R}^{m \times n}$. Then, the value of (4.23) is non-increasing in the prediction horizon length N used.*

Loosely speaking, the result follows from the fact that due to the linear terminal control law, predicted input sequences in (4.23) can always be extended at the end via extending $\mathcal{K}$ suitably.

Proof. Let $\mathcal{K}$ and β_r be part of a solution of LP (4.23). Applying the input sequence $U = \mathcal{K} \frac{1}{\beta_r} x_r$ from the initial condition $\frac{1}{\beta_r} x_r$, the final predicted state $x_{r,N}$ lies within the terminal set and $x_{r,N} = (\hat{A} + \hat{B}\mathcal{K}) \frac{1}{\beta_r} x_r$ holds for suitable matrices $\hat{A} \in \mathbb{R}^{n \times n}$ and $\hat{B} \in \mathbb{R}^{n \times mN}$. An $N + 1$-step feasible input sequence for the same initial condition is given by $U_{N+1} = [U^\top, (K_T x_{r,N})^\top]^\top$. This can be expressed as $U_{N+1} = \mathcal{K}_{N+1} \frac{1}{\beta_r} x_r$ with

$$\mathcal{K}_{N+1} = \begin{bmatrix} \mathcal{K} \\ K_T(\hat{A} + \hat{B}K) \end{bmatrix}.$$

As the matrix $\mathcal{K}_{N+1}$ is independent of the considered state, this applies to all states x_r used in the linear program. Thus, (4.23) formulated for prediction horizon $N + 1$ is feasible with $\mathcal{K}_{N+1}$ and the same β_r values. □

4.3.4 Illustrative example

In the following, we apply the three approaches above to find feasible parametrizations to a numerical example illustrating the methods and their particularities. A more detailed evaluation will be provided below in Subsection 4.5.

Example 4.17 (2D example taken from (Li et al., 2013)). *We consider an example taken from (Li et al., 2013) where also a parametrized MPC scheme is proposed for the same problem. The system under consideration is given by*

$$x^+ = \begin{bmatrix} 0.8 & 1 \\ -0.2 & 1.1 \end{bmatrix} x + \begin{bmatrix} 0.1 & 0.2 \\ -0.2 & 0.5 \end{bmatrix} u$$

with constrained state $x \in \mathbb{X} = \mathbb{R} \times [-4, 4]$ and input $u \in \mathbb{U} = [-1, 1]^2$. The stage cost $\ell(x, u) = \|x\|_2^2 + \|u\|_2^2$ is used and the terminal ingredients are chosen according to the unconstrained LQR setup. We aim at computing parametrizations such that $\mathcal{X}_f = \mathbb{X}_{10}$ holds, i.e., the desired feasible set is the 10-step controllable set to the terminal set. In order to keep the results for all three approaches comparable, we consider parametrizations without constant part, i.e., we require $a_i = 0$. Below, we re-incorporate the constant part and consider its effect.

- **Approach 1: Enlarging clusters and refinement.** *First, we applied Algorithm 7 to compute feasible parametrizations. The training states were sampled from a regular grid of 20×20 states exactly covering $\mathcal{X}_f$, using those states contained in $\mathcal{X}_f$. The MPC problem was formulated using a prediction horizon of $N = 11$ steps. The set $\mathcal{C}$ was taken as hypercube centered at the origin with twice the edge length of the hypercube defined by four neighboring grid points. The clustering algorithm was run from 20 different initial conditions and the best solution obtained was used. We tested the simple refinement (not addressing control performance) for feasibility using different sets of hyperparameters K, q and limiting the refining iterations to $i_{max} = 20$.*

  The simplest combination of hyperparameters which resulted in a feasible parametrization was $q = 1$, $K = 5$. Figure 4.2 shows the corresponding results in state space: On the left hand side, the boundary of the desired feasible set $\mathcal{X}_f$ is shown (in black) together with the training states x_r (black circles) and the sets $\{x_r\} \oplus \mathcal{C}$ which are colored according to cluster membership of the training data. On the right hand side the resulting sets $\hat{\mathcal{D}}_i$ are shown using the same color coding. Note that all sets are generally overlapping and, thus, are shown slightly transparent. In the figure, we see that the clusters have compact shape, lie symmetrically with respect to the origin and can, via the proposed method, be turned into sets $\hat{\mathcal{D}}_i$ which exactly cover the desired feasible set $\mathcal{X}_f$. The chosen hypercubes $\mathcal{C}$ are largely overlapping in the interior of $\mathcal{X}_f$ but it is clear that hypercubes of at least the size used here are required to cover all of $\mathcal{X}_f$ for example at the left and at the right edge of $\mathcal{X}_f$. In the considered two-dimensional example, the numerical complexity of intersecting those hypercubes which are not completely contained in $\mathcal{X}_f$ with $\mathcal{X}_f$ is still well tractable and also the complexity of the resulting sets $\hat{\mathcal{D}}_i$ is still low.

- **Approach 2: Clustering facets and refinement.** Second, we applied Algorithm 9 to compute feasible parametrizations. The training states were sampled from the same regular grid of 20×20 points as before. Here, only points contained in $\mathcal{X}_f \setminus \mathbb{X}_T$ were used, i.e., states in the terminal set were discarded. This is possible because clustering of the facets completely accounts for feasibility of the parametrizations.

  Again we tested the refinement for feasibility using several combinations of hyperparameters q and K. The simplest feasible parametrization was found for the values $q = 1$, $K = 5$. Figure 4.3 shows the corresponding results: On the left hand side, clustered training states and clustered facets of the desired feasible set (here line segments) are shown colored according to cluster membership. On the right hand side, the resulting sets $\hat{\mathcal{D}}_i$ in state space are shown using the same color coding. The clusters again have reasonable and compact shape and clustering of the facets is well integrated into clustering of the interior states.

- **Approach 3: Feasibility via independent part.** Finally, the third approach to compute feasible parametrizations formulated in Algorithm 10 was applied to the same problem. Linear program (4.23) was solved using therein the vertices of the desired feasible set $\mathcal{X}_f$ and the objective function

$$J = 10^3 \hat{\beta} + \sum_{r=1}^{s} \beta_s \tag{4.27}$$

where the constraint $\hat{\beta} \geq \beta_r$ is added to render $\hat{\beta}$ an upper bound on all β_r. That is, an upper bound on all β_r is minimized with a high weight in addition to minimizing all individual β_r variables.

Here, we used the prediction horizon length N as a tuning parameter to render the linear program feasible with $\beta_r \leq 1$. It turned out that for $N \geq 15$ this was achieved. Figure 4.4 illustrates the result for $N = 15$. Depicted are in red the desired feasible set $\mathcal{X}_f$, in blue the achieved feasible set $\hat{\mathcal{X}}_f$ according to (4.24), in black the rays through the vertices of $\mathcal{X}_f$ and marked with black '+' the scaled vertices $\frac{1}{\beta_r} x_R$. In this case the convex hull of the scaled states $\frac{1}{\beta_r} x_R$ is a good inner approximation of the actual feasible set $\hat{\mathcal{X}}_f$ and obviously the desired feasible set $\mathcal{X}_f$ is contained in both sets. Based on this feasibility ensuring linear state feedback, parametrizations using any values for the hyperparameters K and q can be computed.

Even though computation time is not an issue for this small example, comparing computation times for the three approaches gives a hint on their numerical complexity. Considering the simplest feasible case $K = 5$, $q = 1$, the first approach to compute a feasible parametrization required a total of $17\,$s, the second one $12\,$s and the third one required only $7\,$s.

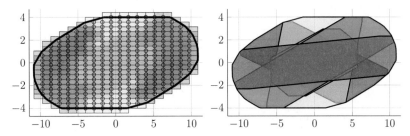

Figure 4.2: Left: Boundary of desired feasible set $\mathcal{X}_f$ (black), training states (black 'o'), sets $\{x_r\} \oplus \mathcal{C}$ colored according to cluster membership of states x_r. Right: Resulting overlapping sets $\hat{\mathcal{D}}_i$, shown slightly transparent.

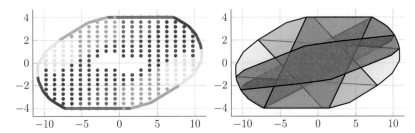

Figure 4.3: Left: Clustered training states and clustered facets of desired feasible set $\mathcal{X}_f$. Right: Resulting overlapping sets $\hat{\mathcal{D}}_i$, shown slightly transparent.

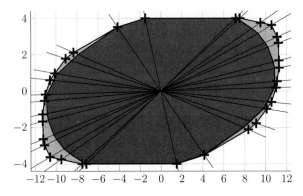

Figure 4.4: Desired feasible set $\mathcal{X}_f$ (red), scaled states $1/\beta_r x_r$ ('+'), rays through states x_r, achieved feasible set $\hat{\mathcal{X}}_f$ (blue).

4.4 Semi-explicit MPC for linear systems: The online algorithm

Whereas the last section was dedicated to the computation of feasible parametrizations in the offline part of the semi-explicit MPC approach for linear systems, the current section discusses the online counterpart. First, we take a quick look at the general method and its properties which are mainly inherited by the generic scheme formulated in Section 3.4. Second, the online solution (more generally evaluation) of the parametrized optimization is discussed. Third, an offline stability test for the parametrization is formulated which allows to employ a simplified online scheme if a parametrization has passed this test.

4.4.1 The basic *online* semi-explicit MPC algorithm for linear systems

Clearly, the general online semi-explicit MPC scheme formulated in Algorithm 4 (Section 3.4.1) is readily applicable to linear MPC problems and all general system theoretic properties which were established in Section 3.4.2 remain valid. In particular, the results on closed-loop asymptotic stability (Theorem 3.6), strong and recursive feasibility (Lemma 3.10 and Proposition 3.11), and the upper bound on the closed-loop costs (Proposition 3.12) remain valid.

Moreover, some properties of linear MPC problems can be exploited to simplify matters further and establish results beyond the latter general ones.

- As stated in Lemma 4.2, the feasible sets of the parametrized mpPs resulting from linear MPC problems are polytopes. Thus, in the online semi-explicit MPC algorithm feasible parametrized mpPs for a given state x can be found via simple polytopic set membership tests.

- Exploiting the structure of different types of parametrized mpPs, different strategies for their numerical solution are applicable and best suitable. This aspect will be discussed more in detail in Section 4.4.2 below.

- Among the general schemes discussed in Section 3.4 were some varieties of the semi-explicit MPC scheme which explicitly use in the optimization the candidate solution obtained by shifting the previous input sequence by one time-step and appending the input obtained from the terminal control law. For linear MPC problems, this approach can render some simplifications inapplicable which would otherwise be possible. We will discuss this together with the latter item below.

- For linear MPC problems with quadratic stage and terminal cost functions, an input determined as linear state feedback is optimal in a region around the origin. In this situation, it is reasonable to employ within the semi-explicit MPC scheme this linear feedback for states in an elliptic set around the origin, i.e, to apply the semi-explicit scheme in a dual mode strategy. Testing whether the state is contained in such elliptic set is numerically cheap compared to the benefit of avoiding any numerical optimization procedure in cases where the state is contained in the set.

- Again for linear MPC problems with quadratic stage and terminal costs, an upper bound on the relative closed-loop cost increase incurred due to the application of the parametrization can be computed. This will be addressed next.

Upper bound on relative closed-loop cost increase

For the following considerations we restrict our attention to control problems with quadratic stage and terminal cost. That is, in this subsection a stage cost $\ell(x, u) = x^\top Q x + u^\top R u$ for given positive definite matrices Q, R and a terminal cost $V_T(x_N) = x_N^\top P_{\text{ARE}} x_N$ with P_{ARE} the positive definite solution of the discrete-time algebraic Riccati equation associated with A, B, Q and R and $\kappa(x) = K_{\text{LQR}} x$ with K_{LQR} the corresponding LQR state feedback gain, is assumed. Our goal is to find an upper bound on the relative closed-loop cost increase

$$\mathcal{P}(x) = \frac{J_{\text{seMPC}}^{\text{cl}}(x)}{J_{\text{nom}}^{\text{cl}}(x)} - 1 \qquad (4.28)$$

caused by the application of the parametrization for initial conditions x in a set $\mathcal{C}$.

Lemma 4.18. *Consider a linear MPC problem with quadratic stage and terminal cost and let $\mathcal{C}$ be a polytope with vertices $\{v_1, \ldots, v_{r_v}\}$ such that $\mathcal{C} \subseteq \mathcal{D}_i \subset \mathcal{X}_f$ for some i and $0 \notin \mathcal{C}$. Applying Algorithm 4 it holds that*

$$\mathcal{P}(x) \leq \frac{k_1}{\tau k_2} - 1 \quad \text{for } x \in \mathcal{C} \qquad (4.29)$$

with $k_1 = \max_{j \in \{1, \ldots, r_v\}} V_{pi}^*(v_j)$ and $k_2 = \min_{x \in \mathcal{C}} J_{nom}^{cl}(x)$.

Proof. According to Proposition 3.12, we have that $J_{\text{seMPC}}^{cl}(x) \leq \frac{1}{\tau} V_{pi}^*(x)$. The non-parametrized mpQP is jointly convex in x and in U (Bemporad et al., 2002), thus, the objective function of $\mathbf{P}_{p_i}(x)$ is jointly convex in x and in $\tilde{U}$ (Lemma 4.3) and, thus, $V_{pi}^*(\cdot)$ is convex (Theorem 2.8). Hence, it holds for $x \in \mathcal{C}$ that $V_{pi}^*(x) \leq \max_{x \in \mathcal{C}} V_{pi}^*(x) = k_1$. Furthermore, $0 \notin \mathcal{C}$ ensures that $k_2 > 0$. $\qquad\square$

For given $\mathcal{C}$, k_2 can be lower bounded by taking such sufficiently long horizon N, formulating a corresponding finite horizon open-loop optimal control problem for this horizon length and solving it with $x \in \mathcal{C}$ as an additional decision variable. This is possible as the problem is jointly convex in x and in U and due to the following: There exists a minimum horizon length $N_{\min}$ such that in the finite horizon open-loop optimal control problem (according to (4.2) in the nominal MPC formulation) for all states x in $\mathcal{C}$ the terminal constraints do not affect the minimum or the minimizer if horizon $N \geq N_{\min}$ is used. The resulting objective value is a lower bound on $J_{\text{nom}}^{cl}(x)$. Details regarding the question for which sets $\mathcal{C}$ a finite minimum horizon length $N_{\min}$ exists such that the terminal constraints become non-relevant for horizons $N \geq N_{\min}$ are found in (Schulze Darup and Cannon, 2016). The bottom line is that at least for each point in the interior of the feasible set for horizon $N \to \infty$ such finite $N_{\min}$ exists and, thus, for each compact set in the interior a finite $N_{\min}$ exists.

The benefit of having the upper bound on $\mathcal{P}(x)$ is threefold. First, it states a systematic way to establish strict guarantees on average and worst case control performance loss for whole sets $\mathcal{C}$. Second, it justifies the refinement for the parametrization based on

(4.13), (4.14) which addresses control performance via optimizing open-loop performance for some sampled states. The refinement in this case addresses minimization of a simplified approximation of $\frac{k_1}{k_2}$. Beyond that, the result could be used to optimize the choice of training states in the interior of $\mathcal{X}_f$ for which control performance is explicitly addressed in the refinement. Loosely speaking, if the states are chosen such that considering them as vertices of sets $\mathcal{C}$ for which good upper bounds on $\mathcal{P}(\cdot)$ are obtained via (4.29), the objective of the refinement becomes more meaningful. In particular, this is the case if the values of $J_{\text{nom}}^{\text{cl}}(\cdot)$ in the interior and at the vertices of $\mathcal{C}$ do not differ too much.

4.4.2 Solving the parametrized optimization problem

So far, the core steps of the online semi-explicit MPC Algorithm were formulated in a rather generic way: *Find i^* such that $x \in \hat{\mathcal{D}}_i$, solve the corresponding $\boldsymbol{P}_{p_{i^*}}(x)$.* Clearly, this captures the main idea and simplification of determining a predicted input sequence in the semi-explicit approach. Yet, first of all in the linear case, further considerations on the details of these steps are appropriate. This issue will be addressed next, first with focus on the general procedure and second addressing the numerical online optimization more in detail.

General approach

In the linear case, the sets $\hat{\mathcal{D}}_i$ of states for which the parametrized optimization $\boldsymbol{P}_{p_i}$ is feasible, are generally overlapping polytopes. This introduces flexibility in the online scheme regarding which one out of the feasible optimization problems to solve and whether or not to even solve several ones if possible. To keep the online scheme as simple as possible, it is reasonable to solve only one out of the feasible problems in each time step. As the sets $\hat{\mathcal{D}}_i$ are connected and typically rather large (resulting from typically low numbers K of different sets used), it will frequently happen that in two consecutive time steps the same part of the parametrization remains feasible. So the simplest strategy would be to first test in each time step if the previously used optimization $\boldsymbol{P}_{p_i}$ remained feasible and if so to use it again. Only if it is not feasible, further polytopic set membership tests are used to identify another feasible part of the parametrization. Investing slightly more online computational effort, one could first identify all feasible parametrized optimization problems, then try to identify (using heuristics with little additional computational effort) the one yielding best control performance among the feasible ones and to only solve this optimization. Identifying the best part of the parametrization could be done for example via application of a multiclass support vector machine (Hsu and Lin, 2002) which has been trained to identify for a given state x the index i yielding best control performance for this state. Aiming at improving control performance at the price of increased online complexity, clearly several feasible parametrized optimization problems could be solved online in each time step so that the one resulting in the lowest objective function value can be used to compute the next control input. This could be done in a sequential manner or in a parallel fashion as suggested in (Longo et al., 2011a,b) in the context of move-blocking for different blocking strategies. This approach would in fact mean to also consider the index i as an additional decision variable yielding overall a mixed integer optimization problem. Beyond that, in the case of a scalar decision variable considered in detail below in Chapter 5, solving the optimization is almost as simple as detecting feasibility. So, in this case it is reasonable to solve all optimization problems which have been identified to be feasible at very little additional computational costs.

In order to solve the parametrized optimization problems, different strategies are applicable. Clearly, the generic and generally suitable approach is to apply a numerical optimization strategy online which benefits from application of the parametrizations. More detailed considerations on that are given below. In the most common cases of having quadratic or linear costs in the MPC formulation, explicit strategies are applicable as well. As discussed above, applying the parametrizations for linear systems to an mpQP or an mpLP, again an mpQP or an mpLP, respectively, results. As in the non-parametrized case, explicit solutions to these problems could be pre-computed offline and applied online. The benefit of applying the parametrization in this case is a reduced number of critical regions the solutions comprise. Exploiting that also these explicit solutions are overlapping in state space, they could be simplified by removing redundant critical regions. Taking this idea further and combining it with the above mentioned strategy to solve all feasible parametrized optimization problems and taking the best solution, a multi-parametric mixed-integer linear or quadratic program would be obtained. In principle, these problems are amenable to explicit solution strategies as well (Oberdieck and Pistikopoulos, 2015). An intermediate strategy, conceptually similar to simplifying explicit solutions by removing overlapping regions would be to resort to using so-called "envelopes of solutions" defined in the latter reference. An alternative intermediate strategy is the almost-explicit approach applicable to cases with scalar decision variable discussed in Chapter 5 of this thesis.

How numerical optimization benefits from using the parametrizations

Next, we examine the numerical solution of the parametrized optimization more in detail. Interestingly, hardly any detailed evaluation specifically of the numerical complexity of solving a parametrized MPC optimization problem is available. This might be partly due to the diversity of available numerical solution strategies which render a general and at the same time detailed evaluation of the benefits impracticable. Nevertheless, there seems to be a consensus that parametrizations simplify numerical solution of the optimization problem considering the large number of available theoretical results on simplifying parametrizations in linear MPC schemes as well as the fact that frequently in applications the decision variable is parametrized.

Let us first take a look at the effect of applying a parametrization of the proposed type on a structural level. In the proposed approach, the dimension of the decision variable can be chosen freely as long as feasibility of the corresponding parametrization can be achieved. In numerical examples, typically this was the case for $q = 2$ and in many cases even the choice $q = 1$ turned out to be sufficient. So generally very low-dimensional decision variables are possible. Once a part p_i of the parametrization is plugged into the multi-parametric program, typically a relevant fraction of the constraints therein can be identified to be redundant for all values of the state and thus can be removed. Due to the reduced dimension of the decision variable and the reduced number of constraints, the maximum possible number of different combinations of active constraints is reduced and typically also the number of actually optimal active sets is reduced considerably. These three effects of applying a parametrization (reduced dimension of decision variable, reduced number of constraints, less different optimal active sets) simplify the typically applied numerical optimization procedures.

Let us next look more in detail at the case of applying interior point methods to solve (parametrized) quadratic programs. Within the iterations of an interior point method, the numerically most expensive part is typically the solution of a linear system of equations to find primal (and dual) search directions (Borrelli et al., 2015; Boyd and Vandenberghe, 2009; Wang and Boyd, 2010). If the condensed formulation of the optimization problem is used and generic methods are applied to solve the equations, this operation has cubic complexity in the dimension of the unknowns, i.e., for the general non-parametrized problem this is $\mathcal{O}(m^3 N^3)$. For longer prediction horizons, it can be beneficial if the sparse (non-condensed) problem formulation is used instead and sparsity patterns in the corresponding system of equations are exploited. A complexity of order $\mathcal{O}(N(n+m)^3)$ can then be achieved, see (Wang and Boyd, 2010).[2] Applying a parametrization, however, all sparsity structure the problem might have is lost and the condensed formulation becomes the most suitable representation. In this case, the complexity to solve the equations is of order $\mathcal{O}(q^3)$. Thus, asymptotically, the following holds: Solving the equation is simplified by the parametrization if $q < mN$ using the dense formulation and having $q < \sqrt[3]{N}(n+m)$ hints that the condensed parametrized optimization has advantages over the non-parametrized sparse optimization. The first relation is guaranteed to hold whenever a reasonable parametrization is used. Considering the very low values of q which the proposed method allows to use, also the second relation will generally hold.

A similar argument applies as well if active set methods are used to solve QPs. Therein, the main computational effort of each iteration is also caused by solving a system of linear equations which is considerably reduced in size by the parametrization. Applying a parametrization, beyond that the reduced number of possible active sets might contribute to simplifying and accelerating the procedure.

These considerations are by no means exhaustive and can only serve as general hints on the actual complexity of a particular implementation of the optimization. More detailed investigations on the interplay of using a parametrization and employing a specific, possibly tailored solution approach would be worthwhile. Further following this idea, it would be interesting to see if specific requirements for a parametrization can be formulated which translate into a desired structure of the parametrized optimization. This has some similarities to the approach proposed in (Jerez et al., 2012), where a pre-stabilization-like transformation is employed to generate structure in the condensed problem formulation.

4.4.3 A simplified online scheme enabled by an offline stability test

In all versions of the semi-explicit MPC approach presented so far, a sufficient decrease in the open-loop costs along closed-loop trajectories had to be ensured *online* in order to guarantee closed-loop stability. Either the currently obtained costs were compared to those of the candidate solution and if necessary the candidate solution was applied or, pursuing the same goal, the candidate solution was included into the parametrized optimization. The goal in this subsection is to shift the computational effort associated with this approach offline to further simplify the online procedure. To this end, an offline test for a given

---

[2]Note that also intermediate strategies have been proposed (Axehill, 2015) which can yield minimal complexity if neither of the "extreme" representations is optimal. This is not considered here any further.

parametrization will be formulated such that a parametrization which passes this test is guaranteed to result in an asymptotically stable closed-loop system even if it is applied in a simplified online scheme which does not use the candidate solution at all.

The presented results are applicable to problems with quadratic stage cost $\ell(x, u) = u^\top Q x + u^\top R u$ where Q and R are positive definite matrices and which have a quadratic terminal cost $V_T(x) = x^\top P_{\mathrm{ARE}} x$ where P_{ARE} is the solution of the ARE associated with A, B, Q and R.

In more detail, in the offline procedure for each part p_i of a given parametrization, a stabilizing condition will be verified for states in a set $\mathcal{S}_i \subseteq \hat{\mathcal{D}}_i$. As a result, the simplified semi-explicit MPC scheme formulated in Algorithm 11 can be guaranteed to asymptotically stabilize the origin of the closed-loop system.

Algorithm 11 Simple semi-explicit MPC scheme

Input: mpP (4.2), parametrization p, sets $\mathcal{S}_1, \ldots, \mathcal{S}_K$
1: **loop**
2:     Obtain current state x
3:     Find i^* such that $x \in \mathcal{S}_{i^*}$
4:     Solve $\mathbf{P}_{p_{i^*}}(x)$, get $\bar{U}^*$
5:     Set $U = p_{i^*}(x, \bar{U}^*)$
6:     Define $u_{\mathrm{se}}(x) = [I_m\ 0]U$ and apply $u_{\mathrm{se}}(x)$ to the plant
7: **end loop**

In order to establish a condition which ensures that Algorithm 11 asymptotically stabilizes the origin, we follow an approach which founds on (Scokaert et al., 1999) and has been used in a similar form in previous work, e.g. (Johansen, 2004; Jones and Morari, 2009) and others. The idea is to find a condition which ensures that the value function $V^*(x)$ of the original MPC problem also is a Lyapunov function for the closed-loop system resulting from application of a simplified MPC scheme, here Algorithm 11.

To this end define

$$\overline{V}_i(x) = \min_U U^\top H U + x^\top F U + x^\top Y x$$
$$\text{s.t. } GU \leq Ex + W \tag{4.30}$$
$$u_0 = u_i^p(x),$$

with $\overline{U}$ denoting the corresponding optimizer. Furthermore, $u_i^p(x)$ denotes the first m elements of $p_i(x, \bar{U}^*)$ and, thus, the second constraint in (4.30) requires the first m elements in $\overline{U}$ to equal the first predicted input corresponding to the solution of $\mathbf{P}_{p_i}(x)$. Thus, (4.30) is in fact a bilevel optimization problem as the constraint is generated via a lower-level optimization problem.

Lemma 4.19. *If for a $\gamma \in \mathbb{R}$ the following holds*

$$\overline{V}_i(x) - V^*(x) - u_i^p(x)^\top R u_i^p(x) - \gamma x^\top Q x \leq 0, \tag{4.31}$$

then for $x^+ = Ax + Bu_i^p(x)$ the relation $V^*(x^+) \leq V^*(x) - (1 - \gamma)x^\top Q x$ holds.

Proof. Let $\overline{U}^C$ be the standard candidate input sequence obtained by shifting $\overline{U}$ by one time step and appending the terminal control input. Then we have

$$V^*(x^+) \leq J(x^+, \overline{U}^C) \leq J(x, \overline{U}) - x^\top Q x - u_i^p(x)^\top R u_i^p(x) \leq V^*(x) - (1 - \gamma)x^\top Q x.$$

Therein, the first inequality follows from optimality, the second inequality is due to Assumption 2.3 and the third one follows from (4.31) with $J(x, \overline{U}) = \overline{V}_i(x)$. □

Based thereon, we formulate the following result.

Proposition 4.20. *Let $\gamma \in [0, 1)$ be such that for $i = 1, \ldots, K$ Condition (4.31) is fulfilled for all $x \in \mathcal{S}_i$. Consider $V_\alpha = \{x | V^*(x) \leq \alpha\}$ such that $V_\alpha \subseteq \cup_{i=1}^K \mathcal{S}_i$. Then, for initial conditions $x \in V_\alpha$ Algorithm 11 is recursively feasible, it renders the origin an asymptotically stable equilibrium of the closed-loop system and V_α is contained in its region of attraction.*

Proof. For $\gamma \in [0, 1)$, Lemma 4.19 states that V^* is a Lyapunov function for the closed-loop system if (4.31) is fulfilled. The assumptions of the proposition ensure that (4.31) is fulfilled for $x \in V_\alpha$ independently of the choice of i^* in step 3 of Algorithm 11. Due to the decrease of the Lyapunov function along closed-loop trajectories, V_α is invariant under closed-loop trajectories and, thus, recursive feasibility and recursive fulfillment of condition (4.31) follows. Asymptotic stability follows from V^* being a Lyapunov function.

□

Here, in fact the optimization is recursively feasible but the MPC scheme based on Algorithm 11 is not strongly feasible as V_α is invariant only under optimal parametrized input sequences. The region of attraction can be extended to $\mathcal{X}_f$ if a suitable control strategy is applied while $x \notin V_\alpha$. If a sufficient decrease in the open-loop cost along closed-loop trajectories is enforced while $x \notin V_\alpha$, the state can be guaranteed to enter V_α after a finite number of time steps. Such decrease in the open-loop cost can always be achieved via applying shifted previous input sequences.

The remaining challenge now is to verify Condition (4.31) for sets (rather than just single points) in the state space. For this purpose, we upper bound the function $\delta(x) = \overline{V}_i(x) - u_i^p(x)^\top R u_i^p(x)$ by an affine function. This is not trivial as $\overline{V}_i(x)$ is defined via the bilevel optimization problem (4.30) and generally neither $\overline{V}_i(x)$ nor $\delta(x)$ are convex functions[3] in x. Yet, $\delta(x)$ is at least piecewise convex: Considering $\mathbf{P}_{p_i}(x)$ as multi-parametric quadratic program (which it is due to the quadratic stage and terminal cost), we know that there exist critical regions $\mathcal{CR}_j^{\tilde{U}} \subset \hat{\mathcal{D}}_i$ such that for $x \in \mathcal{CR}_j^{\tilde{U}}$ the optimizer of $\mathbf{P}_{p_i}(x)$ is an affine function of the state x:

$$\tilde{U}^*(x) = F_j^{\tilde{U}} x + f_j^{\tilde{U}} \text{ if } x \in \mathcal{CR}_j^{\tilde{U}}$$
$$\mathcal{CR}_j^{\tilde{U}} = \{x | A_j^{\tilde{U}} x \leq b_j^{\tilde{U}}\}, \quad j = 1, \ldots, n_{CR}$$

with $F_j^{\tilde{U}} \in \mathbb{R}^{mN \times n}$, $f_j^{\tilde{U}} \in \mathbb{R}^{mN}$, $A_j^{\tilde{U}} \in \mathbb{R}^{c_j \times n}$ and $b_j^{\tilde{U}} \in \mathbb{R}^{c_j}$.

Lemma 4.21. *Let $\mathcal{CR}_j^{\tilde{U}}$ be a critical region of the mpQP $\mathbf{P}_{p_i}(x)$. For $x \in \mathcal{CR}_j^{\tilde{U}}$ the function $\delta(x) = \overline{V}_i(x) - u_i^p(x) R u_i^p(x)$ is convex.*

The latter statement is apparent from the relation $\overline{V}_i(x) - u_i^p(x)^\top R u_i^p(x) = V_{N-1}^*(x^+) + x^\top Q x$ where V_{N-1}^* denotes V^* for prediction horizon $N - 1$ and using the affine relation of x and x^+ for $x \in \mathcal{CR}_j^{\tilde{U}}$. This fact can be exploited to formulate the following result.

---

[3] Considering that $V^*(x) \leq \overline{V}_i(x) \leq V_{pi}^*(x)$ and $V^*(x)$ and $V_{pi}^*(x)$ are both convex functions in x, one might conjecture that also $\overline{V}_i(x)$ is convex. Nevertheless, this is not generally the case and counterexamples falsifying the conjecture can be found.

Proposition 4.22. *Let* $\{x_1, \ldots, x_{\hat{r}}\}$ *be* $\hat{r}$ *states in the critical region* $\mathcal{CR}_j^{\bar{U}}$ *of mpQP* $\boldsymbol{P}_{p_i}(x)$ *with* $\mathcal{C} = conv(\{x_1, \ldots, x_{\hat{r}}\})$ *and let* $v(x) = H_v x + h_v$ *such that* $\overline{V}_i(x_r) - u_i^p(x_r)^\top R u_i^p(x_r) \leq v(x_r)$ *holds for* $r = 1, \ldots, \hat{r}$. *Let* $\gamma \in [0, 1)$ *and define*

$$\tau = -\min_{x \in \mathcal{C}, \, U \in \mathbb{R}^{mN}} \left\{ J(x, U) - v(x) + \gamma x^\top Q x \right\}$$
$$\text{s.t. } GU \leq Ex + W. \tag{4.32}$$

If $\tau \leq 0$ *holds, Condition (4.31) is fulfilled for all* $x \in \mathcal{C}$.

Proof. Due to convexity, v satisfies $\overline{V}_i(x) - u_i^p(x)^\top R u_i^p(x) \leq v(x)$ for all $x \in \mathcal{C}$. The affine function $v(x)$ is convex in x and $J(x, U)$ is jointly convex in x and U (Bemporad et al., 2002), so (4.32) is a convex QP. Thus we have for all $x \in \mathcal{C}$

$$\overline{V}_i(x) - V^*(x) - u_i^p(x)^\top R u_i^p(x) - \gamma x^\top Q x \leq v(x) - V^*(x) - \gamma x^\top Q x \leq \tau \leq 0.$$

$\square$

A main source of conservatism in the latter proposition is the quality of the upper bound v. Generally, smaller sets $\mathcal{C}$ will result in tighter upper bounds v and, thus, less conservative results. These findings can be turned into an algorithm which aims at offline verifying Condition (4.31) for a given part p_i of a parametrization and a given polytope $\mathcal{V} \subseteq \mathcal{S}_i$. As input data for the algorithm, a list of simplices $\mathcal{L} = (\mathcal{C}_1, \ldots, \mathcal{C}_{\hat{s}})$ is determined, such that each simplex $\mathcal{C}_s$ is contained in one of the critical regions of $\boldsymbol{P}_{p_i}$, i.e. $\mathcal{C}_s \subseteq \mathcal{CR}_j^{\bar{U}}$ for one j, and $\mathcal{V} = \bigcup_s \mathcal{C}_s$ holds. This can be done for example via a Delaunay triangulation of the critical regions of $\boldsymbol{P}_{p_i}$ based on their vertices. The algorithm then yields a list $\mathcal{L}_f$ of simplices such that Condition (4.31) holds (at least) for all x in the union of these simplices. Algorithm 12 summarizes this offline stability check. Therein, v and volume($\mathcal{C}$) are easily computable as $\mathcal{C}$ is a simplex.

Remark 4.23. *Algorithm 12 typically yields more conservative results if the appearing simplices are close to being degenerated having at least one edge which is large with respect to the volume of the simplex. Thus, this should be taken into account when subdividing simplices to keep the results less conservative. Having this in mind, alternatively the maximum edge length of a simplex could be taken as a criterion to decide in case of infeasibility whether to subdivide the simplex further or to consider it finally infeasible.*

Remark 4.24. *Note that QP (4.32) cannot result in* $\tau \leq 0$ *for any neighborhood of the origin as this would require to bound a purely quadratic function from below using a linear function which is not identically zero. Clearly, this is not possible in any neighborhood of zero. Instead, consider* $V_\beta = \{x | x^\top P_{ARE} x \leq \beta\}$ *and* β *such that* V_β *is completely contained in the terminal region and in the critical region of* $\boldsymbol{P}_{p_i}$ *which contains the origin. Then, for* $x \in V_\beta$, $J^*(x) = x^\top P_{ARE} x$ *holds and the Lyapunov decrease condition can be checked explicitly for the closed-loop system.*

Based on the latter remark and the output of Algorithm 12, the sets $\mathcal{S}_i$ can then be defined. It is straight forward to take $\mathcal{S}_i = \hat{\mathcal{D}}_i$ if $\hat{\mathcal{D}}_i = \{\mathcal{C}_i \in \mathcal{L}_f\}$. Generally, this will not be the case and the list of infeasible simplices in the output of Algorithm 12 will be non-empty. In such situation, defining $\mathcal{S}_i$ naively via the list of feasible simplices would

---

Algorithm 12 Offline stability check

Input: $\mathbf{P}_{p_i}$, QP (4.30), QP (4.32), $\gamma \in [0,1)$, simplex list $\mathcal{L} = (\mathcal{C}_1, \ldots, \mathcal{C}_{\hat{s}})$, minimum volume vol_{min}.

1: **while** $\mathcal{L} \neq \varnothing$ **do**
2:   compute v for first element $\mathcal{C}$ of $\mathcal{L}$
3:   solve QP (4.32) for $\mathcal{C}$
4:   **if** $\tau \leq 0$ **then**
5:     remove set $\mathcal{C}$ from list $\mathcal{L}$
6:     add $\mathcal{C}$ to list of feasible simplices $\mathcal{L}_f$
7:   **else**
8:     **if** volume($\mathcal{C}$) $< vol_{min}$ **then**
9:       add $\mathcal{C}$ to list of infeasible simplices $\mathcal{L}_{inf}$
10:    **else**
11:      subdivide set $\mathcal{C}$ into simplices $\{\mathcal{C}_a, \mathcal{C}_b, \ldots\}$ such that $\mathcal{C} \subseteq \mathcal{C}_a \cup \mathcal{C}_b \cup \ldots$
12:      remove $\mathcal{C}$ from list $\mathcal{L}$
13:      add $\mathcal{C}_a, \mathcal{C}_b, \ldots$ to list $\mathcal{L}$
14:    **end if**
15:   **end if**
16: **end while**
Output: Lists $\mathcal{L}_{inf}$ and $\mathcal{L}_f$

---

generally yield highly complex sets $\mathcal{S}_i$ which contradicts the desired simplification of the online scheme. In this case, an explicit representation of a set $\mathcal{S}_i$ could be obtained as $\mathcal{S}_i = \hat{\mathcal{D}}_i \setminus \mathcal{E}_i$ where $\mathcal{E}_i$ is a suitably chosen polytopic or elliptic set containing the infeasible simplices. Furthermore, implicit representations of the sets $\mathcal{S}_i$ can be applicable. For example, testing in the online algorithm set membership of the current state to the sets $\hat{\mathcal{D}}_i$ in the right order and using the first feasible one can be sufficient to ensure that always a stabilizing input is selected. After a suitable reordering of the sets $\hat{\mathcal{D}}_i$, this would mean to use the definition $\mathcal{S}_i = \hat{\mathcal{D}}_i \setminus \hat{\mathcal{D}}_{i-1} \setminus \cdots \setminus \hat{\mathcal{D}}_1$. This approach is illustrated in a simple example below. Finally, it is reasonable to expect that among several feasible parts p_i of a parametrization the one resulting in the lowest objective function value of the parametrized optimization is stabilizing. Offline, this property could be verified by solving for each of the simplices infeasible in Algorithm 12 for p_i an additional test which verifies that there is a p_j which is stabilizing and yields better open-loop performance on the simplex. Online, several parametrized optimization problems would be solved and the one resulting in best control performance would be used.

Remark 4.25. *For the results in this section, the weighting matrix Q in the stage cost function was required to be positive definite and this property was exploited in the stability proof. This condition could be relaxed to requiring Q to be positive semi-definite with $Q = C^\top C$ for a matrix C such that (A, C) is detectable and in addition exchanging (4.31) for*

$$\overline{V}_i(x) - V^*(x) - \overline{\gamma} u_i^p(x)^\top R u_i^p(x) - \gamma x^\top Q x \leq 0 \qquad (4.33)$$

with $\gamma, \overline{\gamma} \in [0,1)$. Asymptotic stability can in this case be shown based on LaSalle's invariance principle. Condition (4.33) is verifiable completely along the same lines as *presented above for (4.31) as $u_i^p(x)$ is available explicitly depending on x. Clearly, increasing $\overline{\gamma}$ renders (4.33) less conservative so for Q positive definite, (4.31) is to be preferred for (4.33).*

In the literature, there are other methods available for checking closed-loop asymptotic stability of MPC schemes which might seem to be applicable in the situation considered here as well. For example, it is a well established procedure to compute for a given system a simplified explicit controller by neglecting some of the stability ensuring ingredients (i.e. terminal constraints, long prediction horizon, etc.) and to test a posteriori (but still offline) if the resulting controller is asymptotically stabilizing, nevertheless. See for example the corresponding methods provided in the current and the previous version of the Multi-Parametric Toolbox (Herceg et al., 2013; Kvasnica et al., 2004). For this approach it is exploited that the resulting closed-loop system has a piecewise affine representation and that for this system class efficient stability tests are available, see for example (Biswas et al., 2005).

In principle, such an approach is applicable here, as well. But, to this end it is necessary to have a priori offline available a piecewise affine representation of the closed-loop system for which, in addition, the affine parts are defined on polytopes in the state space. This, in turn, would require to define a priori off-line for each part p_i of the parametrization one or several polytopes of states for which it is to be applied in the online optimization and to ensure that these polytopes altogether form a partition of the overall feasible set. In particular this would spoil all freedom in the online algorithm to select the part p_i freely among several feasible ones. For example reusing the one from the previous time step if it remained feasible or to solve several parametrized optimization problems to choose the one with lowest objective function value would not be possible (recall that comparing piecewise quadratic functions would result in a non-polytopic partition). In other words, the advantage of the approach of using $V^*(x)$ as Lyapunov function candidate is that automatically for several parts p_i of a parametrization the Lyapunov decrease condition is verified if possible, conceptually similar to the approach of finding a common Lyapunov function for several closed-loop systems.

For this result, main ingredients were convexity properties of the mpP and existence of an explicit solution of the mpP. Both properties are also fulfilled if the mpP is a multi-parametric linear program instead of a multi-parametric quadratic program. Thus, corresponding results could as well be formulated for MPC problems with linear or infinity norm stage and terminal costs.

Illustrative example

Next, we briefly demonstrate application of the offline stability test and reconsider for this purpose the previous illustrative example.

Example 4.26 (Example 4.17 revisited). *A feasible parametrization was computed applying the second approach and using the values $K = 2$, $q = 2$ and reusing all remaining parameters as given above. Then the explicit solution of both parametrized mpQPs was computed and, finally, Algorithm 12 and Remark 4.24 were employed to verify Condition (4.31) for $\gamma = 0.9999$. The results are illustrated in Figure 4.5 and 4.6. For part p_1 of the parametrization, the condition could be verified for all $x \in \hat{\mathcal{D}}_1$ (union of red and yellow sets in right part of Figure 4.5). For part p_2, the condition could be verified only for a* *subset of $\hat{\mathcal{D}}_2$ (union of green and yellow sets in right part of Figure 4.5). The red area contained in between the large yellow sets is the union of the simplices found to be infeasible in Algorithm 12. These simplices are all in volume below the threshold used in the algorithm, except for those simplices for which Remark 4.24 was applied. It is interesting to see that these infeasible simplices are all located away from the states which have been assigned to the respective cluster (green points and line segments in left part of Figure 4.5). This supports the interpretation that each part p_i of the parametrization is actually tailored for a specific region in state space and that this tailoring might make this part unsuitable for other states. We furthermore note that all states for which Condition (4.31) could not be verified for p_2 are contained in $\hat{\mathcal{D}}_1$. Thus, a stabilizing online strategy would be to first test if p_1 is feasible for the current state and apply it if possible and to use p_2 only if this is not the case. This corresponds to defining the sets $\mathcal{S}_1 = \hat{\mathcal{D}}_1$ and $\mathcal{S}_2 = \hat{\mathcal{D}}_2 \setminus \hat{\mathcal{D}}_1$. In Figure 4.6, the data generated while running Algorithm 12 for the first cluster is illustrated including simplices generated (red), affine functions $v(x)$ (blue) and sets for which Remark 4.24 was used to show stability (cyan).*

4.5 Examples and evaluation

Next, we evaluate the semi-explicit MPC approach for linear systems in numerical examples and then provide a theoretical evaluation, also taking into account the findings in the examples.

4.5.1 Numerical examples

For the numerical examples to be considered next, we restrict ourselves to parametrizations without constant part, i.e., with $a_i = 0$. This choice will be justified below where we evaluate the effect of using a constant part. We first re-consider Example 4.17 once more, this time with a focus on the effect of the hyperparameters K and q on feasibility and control performance and we will evaluate the importance of the candidate input sequences in the online algorithm.

Example 4.27 (Example 4.17 revisited). *We applied the second method to compute feasible parametrizations (Algorithm 9) using values $K \in \{1, 2, \ldots, 7\}$ and $q \in \{1, 2, 3, 4\}$. In the refinement, the performance enhancing objective functions were used and after feasibility had been achieved, two more iterations were executed in order to improve control performance of the resulting parametrizations. The control problem and all remaining parameters were left unchanged and used as introduced above.*

Whereas the refinement procedure was not for all combinations of K and q feasible, for each $q \geq 1$ a sufficient value for K was found to achieve feasibility. Figure 4.7 illustrates for different values of q the lowest value of such K (black dots). Obviously, K and q can be traded off against each other. Using a scalar decision variable ($q = 1$), five different parts are required in the parametrization ($K = 5$), whereas in the other extreme case, using a four-dimensional decision variable ($q = 4$) even a single part ($K = 1$) is sufficient. In (Li et al., 2013) a global parametrization is presented which essentially recovers the same feasible set as obtained here, but, which uses a five-dimensional decision variable. The corresponding combination is marked in Figure 4.7 as well (gray diagonal cross).

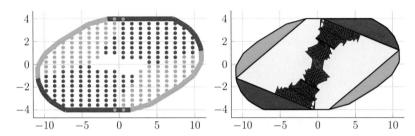

Figure 4.5: Left: Clustered states and facets. Right: States for which Condition (4.31) has been verified for $\gamma = 0.9999$ (red/green: condition verified only for cluster corresponding to respective color, yellow: condition verified for both clusters).

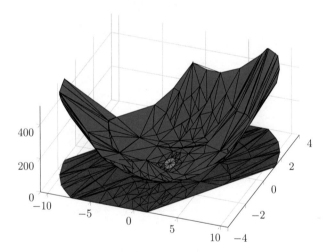

Figure 4.6: Data generated while running Algorithm 12 corresponding to the red cluster in Figure 4.5 (red: simplices generated while running algorithm, blue: affine mappings $v(x)$, cyan: cost decrease verified via Remark 4.24).

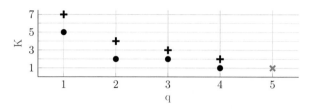

Figure 4.7: For several values of q lowest value of K which achieves same feasible set as reported in (Li et al., 2013) (black dots) and same control performance (black cross). In (Li et al., 2013) one globally defined parametrization with $q = 5$ is used (gray diagonal cross).

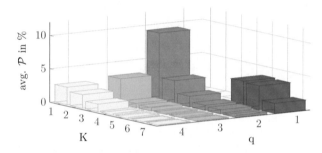

Figure 4.8: Average closed-loop performance loss over 1 896 initial conditions for different values of K and q for Example 4.17.

In order to evaluate control performance, we executed closed-loop simulations from 1 896 initial conditions sampled from a regular grid covering $\mathcal{X}_f \setminus \mathbb{X}_T$ applying Algorithm 4. For different combinations of values for K and q, the suboptimality $\mathcal{P}(x_0) = J^{cl}_{param}(x_0)/J^{cl}_{nom}(x_0) - 1$ averaged over all initial conditions was evaluated. The results are summarized in Figure 4.8. A very clear tendency is obvious that increasing either of the hyperparameters K or q improves closed-loop performance. To achieve a certain level of performance, several combinations of values for K and q are suitable such that a tradeoff and flexibility in the design are possible. Only few single outliers are observed where an increase in K or q results in slightly worse closed-loop performance. Again the results are compared to those reported in (Li et al., 2013) using a five-dimensional decision variable. Therein, an average closed-loop suboptimality of $\mathcal{P} = 2.13\%$ was reported executing closed-loop simulation from a number of initial conditions. Applying the semi-explicit MPC approach, for each $q \in \{1, 2, 3, 4\}$ a value for K could be found such that on average at least the same control performance was observed. For each value of q, in Figure 4.7 the lowest value of such K is marked (black cross). It is obvious that due to the piecewise definition of the parametrization in a semi-explicit MPC approach a lower-dimensional decision variable can be used recovering similar control performance.

Next, we take a look at the relevance of candidate solutions in the online procedure given in Algorithm 4. Applying Lemma 3.10, we know that initialized feasible, the parametrized *optimization is guaranteed to remain feasible for all future time-steps. Thus, application of the candidate solution can never become necessary because of infeasibility and, instead, it might be necessary only to ensure asymptotic stability of the closed loop via a sufficient open-loop cost decrease, $J(x,U) \leq J_{comp}$. Choosing the parameter τ in the online algorithm to be a small positive constant, the candidate input sequence was only required for the parametrization with $K = 2$, $q = 2$ at all. For this parametrization in line 8 of Algorithm 4 $J_{comp} < J(x,U)$ occurred for 2.5 % of all states visited during the simulations. Overall, we conclude that for this example the cost comparison in line 8 of Algorithm 4 is mainly required to establish theoretical guarantees. As is to be expected, choosing τ closer to 1, the frequency with which $J_{comp} < J(x,U)$ occurred increased.*

Next, we consider a slightly larger example which has been addressed in (Summers et al., 2011) in a suboptimal explicit MPC context before. The goal is to show that offline computations remain well tractable for a four-dimensional system with rather long prediction horizon, a tradeoff of control performance against values of K and q as observed in the first example is possible and that achievable control performance is high even using just scalar decision variables. A brief comparison to the results reported in (Summers et al., 2011) will be drawn below.

Example 4.28 (Four-dimensional example taken from (Summers et al., 2011)). *The system under consideration is given by the dynamics*

$$x^+ = \begin{bmatrix} 0.4035 & 0.3704 & 0.2935 & -0.7258 \\ -0.2114 & 0.6405 & -0.6717 & -0.0420 \\ 0.8368 & 0.0175 & -0.2806 & 0.3808 \\ -0.0724 & 0.6001 & 0.5552 & 0.4919 \end{bmatrix} x + \begin{bmatrix} 1.6124 \\ 0.4086 \\ -1.4512 \\ -0.6761 \end{bmatrix} u$$

and the state constraints $\|x\|_\infty \leq 5$ and input constraints $|u| \leq 0.2$. The stage cost is $\ell(x,u) = \|x\|_2^2 + 0.2u^2$ and the terminal cost, terminal controller and terminal constraint set corresponding to the unconstrained LQR setup are used. A prediction horizon of $N = 17$ is implemented. To compute a desired feasible set, 256 points were sampled from the facets of $\mathcal{X}_{12}$, the 12-step controllable set to the terminal set, and $\mathcal{X}_f$ was taken as their convex hull.

For different combinations of K and q we computed two types of feasible parametrizations applying the second approach (Algorithm 9). For the first type we aimed at low offline computation times to find feasible parametrizations. This was achieved using $n_a = 650$ training states sampled from a regular grid and employing in the refinement the objective function which only addresses feasibility. Computation of the desired feasible set and the training data set required in total 25 s. It turned out that for $q = 1$ the smallest feasible value for K was $K = 7$. For $q = 2$ the refinement for $K = 2$ was feasible. In this range parametrizations for different values of K and q could then be computed in at most 30 s. Then for the same values of K and q parametrizations were computed investing higher offline computation times and aiming at achieving higher control performance. In this case $n_a = 1562$ training states were sampled and in the refinement the objective addressing control performance was used. Offline computations were still well tractable and required less than 400 s in all cases. All offline computation times for the first case are given in Table 4.1 and for the second case in Table 4.2.

In closed-loop simulations from 10^5 initial conditions sampled randomly from $\mathcal{X}_f \setminus \mathbb{X}_T$ we evaluated control performance. In fact the parametrizations computed via the more expensive

Table 4.1: Comparison of parametrizations computed using different values of K and q aiming at quick offline computations for Example 4.28.

	$q = 1$		$q = 2$	
criterion	$K = 7$	$K = 15$	$K = 2$	$K = 10$
offline computation time	26 s	30 s	17 s	24 s
average relative suboptimality $\mathcal{P}_{avg}$	14.8 %	12.1 %	12.9 %	11.5 %

Table 4.2: Comparison of parametrizations computed using different values of K and q aiming at high control performance for Example 4.28.

	$q = 1$		$q = 2$	
criterion	$K = 7$	$K = 15$	$K = 2$	$K = 10$
offline computation time	311 s	294 s	392 s	331 s
average relative suboptimality $\mathcal{P}_{avg}$	11.8 %	6.3 %	5.0 %	3.1 %
average numb. of constraints of $P2_i$	50.6	42.4	58.0	46.2
$\sum_i \#$ crit. regions of expl. sol. of $P2_i$	254	458	472	1 090

offline procedure performed better. As in the previous example, control performance was clearly improved by increasing either K or q. The best control performance observed was only 3.1 % suboptimal on average. All values are included in Tables 4.1 and 4.2, respectively.

Next, we again take a look at the significance of the candidate input sequences in the online scheme. Over all simulations, the parametrized optimization never became infeasible. For the quickly computed parametrization using $K = 2$, $q = 2$, the largest number of cases was observed in which a candidate sequence was needed to achieve a decrease in the predicted costs (3.7 % of all states visited during simulations). For all other considered parametrizations this was even less and none of the parametrizations computed aiming at high control performance required application of the candidate input sequence.

In order to facilitate evaluation of the complexity of solving the parametrized optimization problems $\boldsymbol{P}_{p_i}$ online, we state further characteristics thereof next. Besides the dimension of the decision variable, as discussed above the number of non-redundant constraints can have an impact on the computational effort solving the optimization. The original non-parametrized mpQP comprises 82 non-redundant constraints whereas in the parametrized problems $\boldsymbol{P}_{p_i}$ for different combinations of K and q on average 42.4 – 58.0 non-redundant constraints were present. Thus, in all cases a significant reduction could be achieved. Furthermore we determined the explicit solutions of all multi-parametric quadratic programs $\boldsymbol{P}_{p_i}$. Depending on the considered combination of K and q, 254 – 1 090 critical regions were required to describe the complete solution, i.e. summing the number of critical regions over all parts i for a considered combination of K and q. Considering that the explicit solution of the original non-parametrized mpQP consists of more than 50 000 critical regions (Summers et al., 2011), again a drastic simplification could be achieved. The number of constraints and critical regions are given in the next-to-last and last row of Table 4.2, respectively.

Finally, we set the current results into relation with the results reported in (Summers et al., 2011). The control performance reported in (Summers et al., 2011) for a simplified

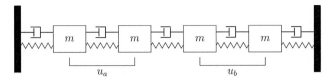

Figure 4.9: System of 4 oscillating masses. The dark blocks on the side represent walls, u_a and u_b are forces acting between the masses.

explicit approach applied to the same example is on average "less than 3 %" suboptimal. The performance for the proposed method using the values $q = 2$, $K = 10$ is in a similar range. The online computational requirements of both methods are not directly comparable as the multiresolution approximation proposed in (Summers et al., 2011) can be evaluated in a particularly efficient way. Nevertheless, the fact that the solution in (Summers et al., 2011) required 3 633 "hierarchical details" and the solution reported here requires only 1090 critical regions might be seen as an indicator that the online complexity of both solutions is in a similar order of magnitude. Furthermore, the number of critical regions of the simplest solution found here amounts to less than 7 % of the number of hierarchical details of the solution reported in (Summers et al., 2011).

Summarizing, in this example the second method to compute feasible parametrizations could be well applied to a four-dimensional example problem. Offline computation times to obtain a feasible parametrization were low and investing moderate offline computation times, parametrizations could be computed which achieve high closed-loop control performance. Whereas computing and using the exact explicit solution of the original problem seems impracticable in this case, computing explicit solutions for the parametrized problems was well possible and the complexity of the solutions was low. Control performance achieved and the complexity of these explicit solutions was comparable to the referenced suboptimal explicit solution.

Finally, we consider an example which is motivated by the control of four oscillating masses. This type of problems serves as a kind of benchmark problem for fast MPC algorithms and has been considered among others in (Jerez et al., 2014; Wang and Boyd, 2010; Zeilinger et al., 2011).

Example 4.29 (Oscillating masses example). *In this example, we consider a system of four masses which are interconnected by spring-damper systems of which two are connected to walls at the sides and with two forces each acting between two of the masses, see Figure 4.9 for an illustration. The masses have value 1, the spring constants are 1 and the damping constants are 0.5. The resulting continuous time system is discretized in time using a sampling time of $0.5\,s$. This yields a system of the form $x^+ = Ax + Bu$ with matrices*

$$A = \begin{bmatrix} 0.8039 & 0.0839 & 0.0086 & 0.0006 & 0.3670 & 0.0576 & 0.0054 & 0.0004 \\ 0.0839 & 0.8125 & 0.0845 & 0.0086 & 0.0576 & 0.3724 & 0.0579 & 0.0054 \\ 0.0086 & 0.0845 & 0.8125 & 0.0839 & 0.0054 & 0.0579 & 0.3724 & 0.0576 \\ 0.0006 & 0.0086 & 0.0839 & 0.8039 & 0.0004 & 0.0054 & 0.0576 & 0.3670 \\ -0.6765 & 0.2573 & 0.0471 & 0.0047 & 0.4657 & 0.2125 & 0.0321 & 0.0030 \\ 0.2573 & -0.6294 & 0.2620 & 0.0471 & 0.2125 & 0.4978 & 0.2154 & 0.0321 \\ 0.0471 & 0.2620 & -0.6294 & 0.2573 & 0.0321 & 0.2154 & 0.4978 & 0.2125 \\ 0.0047 & 0.0471 & 0.2573 & -0.6765 & 0.0030 & 0.0321 & 0.2125 & 0.4657 \end{bmatrix}$$

and

$$B = \begin{bmatrix} 0.0931 & -0.0938 & -0.0093 & -0.0006 & 0.3094 & -0.3149 & -0.0525 & -0.0051 \\ 0.0006 & 0.0093 & 0.0938 & -0.0931 & 0.0051 & 0.0525 & 0.3149 & -0.3094 \end{bmatrix}^{\top}.$$

The states of the system represent positions and velocities of the masses and the inputs u_a and u_b represent forces acting in between the masses, cf. Figure 4.9. The constraints $x \in [-4, 4]^8$ and $u \in [-1, 1]^2$ are imposed. For the MPC setup the stage cost $\ell(x, u) = \|x\|^2 + \|u\|^2$ is used and the terminal cost, terminal controller and terminal constraint set corresponding to the unconstrained LQR setup are chosen.

For this MPC problem the third method to find feasible parametrizations based on Algorithm 10 was applied. As a reference feasible set $\mathcal{X}_f^{ref}$ we used the eight-step controllable set to the terminal set. The set was not computed explicitly, instead we used $s = 1\,500$ representative points sampled from the facets of $\mathcal{X}_f^{ref}$ in LP (4.23). We then employed an MPC setup with a prediction horizon of $N = 13$. In the linear program, an objective function of the form (4.27) as in Example 4.17 was used which addresses in a combined fashion scaling of the individual states and maximization of a common bound on the scaling of the states.

In LP (4.23) for all states $1/\beta_s > 0.95$ was achieved and for more than $77\,\%$ of all states $1/\beta_s \geq 1$ was achieved. That is, the convex hull of the sampled points scaled by the factor 0.95 could be included in the feasible set $\hat{\mathcal{X}}_f$ and 77% of all states sampled from the facets of $\mathcal{X}_f^{ref}$ were included in the feasible set $\hat{\mathcal{X}}_f$.

We restricted our attention to parametrizations with scalar parameter, i.e. $q = 1$, and computed parametrizations for $K \in \{2, 5, 15\}$. Here, we used a weighted version of the clustering (3.8) which weights the first predicted inputs higher and thereby increases approximation accuracy thereof, which in turn results in improved control performance. Computing the first parametrization using $K = 2$ required a total of $59\,s$ (including computation of training data, solution of the LP, clustering). All subsequent parametrizations could be computed even quicker as the result of LP (4.23) and the training data could be reused. See the corresponding row of Table 4.3 for all computation times.

In closed-loop simulations executed from $5\,000$ initial conditions sampled randomly from $\hat{\mathcal{X}}_f \setminus \mathbb{X}_T$ we evaluated control performance. Even using only $K = 2$, control performance was very close to optimal (on average $1.3\,\%$ suboptimal) and suboptimality could be reduced to an average of only $0.3\,\%$ by increasing K to $K = 15$.

In order to give an impression of the online complexity required to obtain a solution of the parametrized optimization problems, we state further characteristics thereof next. The original non-parametrized mpQP comprises 172 non-redundant constraints whereas in the parametrized problems $\boldsymbol{P}_{p_i}$ this number could for all values of K be reduced to an average of approximately 145 non-redundant constraints, see the next-to-last row of Table 4.3 for all values. In this example, the simplification of the explicit solution and the reduction in the number of critical regions therein is much more significant. All parametrized optimization problems $\boldsymbol{P}_{p_i}$ could be solved explicitly, each within a few minutes and the resulting solutions had well tractable complexity. In more detail, depending on the chosen value for K, 214 – 1541 critical regions were obtained, summing the number over all i for a particular parametrization, see the last row of Table 4.3 for all values. In contrast to that, for the original mpQP no explicit solution could be computed within a reasonable amount of time and we terminated the computation prematurely when already thousands of regions had been found after hours of computations.

Table 4.3: Comparison of parametrizations computed using $q = 1$ and different values for K for Example 4.29.
*Note that LP (4.23) was solved only once for the first parametrization ($K = 2$) and the result was then reused. The same holds for the training data.

criterion	$K = 2$	$K = 5$	$K = 15$
offline computation time	59 s	13 s*	58 s*
average relative suboptimality $\mathcal{P}_{avg}$	1.3 %	0.5 %	0.3 %
average numb. of constraints of $P2_i$	145	144.8	144.7
$\sum_i$ # crit. regions of expl. sol. of $P2_i$	214	507	1 541

Summarizing, in this example a semi-explicit MPC scheme was set up and applied for a problem with eight-dimensional state space and 26-dimensional decision variable. Due to these relatively high dimensions, this MPC problem is clearly beyond applicability of exact explicit solutions. In contrast, the third approach to compute a feasible parametrization was well applicable requiring only moderate offline computation times. Controlling the system via the semi-explicit MPC scheme does not only guarantee closed-loop asymptotic stability but in fact yields close to optimal control performance even using only scalar decision variables.

4.5.2 Comparison and evaluation of the methods to compute feasible parametrizations

In the following, we compare the three approaches to compute feasible parametrizations theoretically.

Complexity of the sets $\hat{\mathcal{D}}_i$

Let us start by having a look at the complexity of the sets $\hat{\mathcal{D}}_i$ resulting from the different approaches as this affects several other aspects to be considered below. In the first approach, each training state is mapped to the 2^n vertices of an n-dimensional hypercube and each one of them becomes a potential vertex of one of the sets $\hat{\mathcal{D}}_i$. In the second approach, mainly the training states (besides the vertices of the desired feasible set) play the role of potential vertices of sets $\hat{\mathcal{D}}_i$. Thus, the number of potential vertices in the first approach is about 2^n times the number of potential vertices in the second approach. In addition, intersecting the convex hulls with $\mathcal{X}_f$ in the first approach can create additional vertices. As a consequence, the first approach generally yields more complex sets $\hat{\mathcal{D}}_i$ than the second one and with growing state space dimension n, the difference quickly grows. The third approach allows to choose the sets $\hat{\mathcal{D}}_i$ totally freely within $\hat{\mathcal{X}}_f$, so there the simplest definition is possible. Beyond that, the number of hyperplanes defining the set $\hat{\mathcal{X}}_f$ is a priori bounded as is evident from (4.24).

Numerical complexity and scalability

Next, we consider numerical complexity of the approaches. Criteria of interest in this respect are minimal complexity for a given problem, scalability in the state space dimension and required effort to iterate over different hyperparameters K and q. The complexity of the approaches is mainly affected by the type of operations required therein and by the amount of (training) data used in the approach. In the first approach, for the computation of each set $\hat{\mathcal{D}}_i$ a facet enumeration has to be executed in order to evaluate the intersection of the convex hull of the hypercubes with the desired feasible set $\mathcal{X}_f$ and in order to obtain the vertices of the resulting set $\hat{\mathcal{D}}_i$ also a vertex enumeration has to be executed. Both operations do not scale too well in the problem dimension. Furthermore, in this approach relatively many training states have to be sampled from the whole desired feasible set as otherwise the enlarging hypercubes $\mathcal{C}$ would have to be chosen very large, rendering the approach conservative.

For the second approach, computation of the sets $\hat{\mathcal{D}}_i$ is generally much simpler as only the vertices of the convex hull of a set of points are to be identified and a facet enumeration of the convex hull is computed. The fact that in this case the sets $\hat{\mathcal{D}}_i$ are defined as the convex hull of a much smaller set of points simplifies matters further. Generally less training states are sufficient in this approach as only the vertices of the simplices at the facets are required for feasibility. In this case the complexity of the desired feasible set is decisive for the amount of training data required and, hence, also for the minimal achievable numerical complexity. Thus, the first approach requires some complex operations in order to compute the sets $\hat{\mathcal{D}}_i$ and has due to the required operations and amount of training data limited scalability in the state space dimension. The second approach behaves better in these aspects. In both cases a refinement of the quantities M_i, $\mathcal{K}_i$ and a_i is required. This can be numerically expensive, and complexity of the refinement is affected largely by the complexity of the sets $\hat{\mathcal{D}}_i$, in particular the number of vertices these sets comprise. This means that again the generally simpler sets resulting from the second approach are advantageous. For both approaches iterating over hyperparameters K and q requires to recompute $\hat{\mathcal{D}}_i$ and to re-execute the refinement for each considered combination of K and q. In the second approach, the preparation of the simplices within the training data has to be executed only once and the result can be reused in the iterations. The numerical complexity of the third approach is considerably lower and the required operations scale better in the state space dimension. In this case, only one linear program has to be solved and a hyperplane representation of the set $\hat{\mathcal{X}}_f$ is automatically obtained. The result can be re-used to iterate over different values for K and q such that each iteration causes only little additional numerical costs once the LP has been solved. The overall complexity in this case mainly depends on the number of states used in the linear program. Technically, no additional training data is needed at all beyond these states. In an application, one might use states which are known to appear during operation of the process to be controlled to be rendered feasible in the linear program and as additional training data to improve performance for them. Clearly, larger flexibility in the choice of training data and, first of all, the possibility to use less training states contribute considerably to scalability of the method.

Without further optimizing implementation of the offline algorithms and employing a standard desktop PC, the following guidelines regarding applicability of the three approaches can be given. The first approach to find feasible parametrizations is generally applicable to systems with up to 3–4 states, the second approach is applicable to systems with up to 4–6 states and the third approach is applicable to problems with up to 10 states. Yet, these numbers are highly dependent on the problem details, and first of all, on the complexity of

Table 4.4: Comparison of the strengths of the approaches to find feasible parametrizations.

criterion	1st approach	2nd approach	3rd approach
required training states	many	$\longrightarrow$	few
numerical complexity	significant	$\longrightarrow$	low
scalability	fair	$\longrightarrow$	good
flexibility	large	$\longleftarrow$	limited
feasibility/performance	high	$\longleftarrow$	fair

the desired feasible set $\mathcal{X}_f$. Using a simplex as $\mathcal{X}_f$, its number of vertices and facets would scale only linearly in the state dimension. Thus, for such problem, the second and the third approach to find feasible parametrizations would be applicable far beyond the given values.

Summarizing the latter considerations, the first approach to find feasible parametrizations is numerically most expensive and requires the largest number of training states. The second approach behaves better in both aspects and minimal complexity is defined mainly by the complexity of the desired feasible set. The third approach uses only numerically simple operations and requires a minimum amount of training states. Thus, the third approach scales by far best in the state space dimension, followed by the second approach.

Flexibility of the resulting parametrization

Next, we consider how the different approaches affect flexibility of the sets $\hat{\mathcal{D}}_i$ and of the parametrization in general. This is interesting as it affects the capability of a parametrization of fixed values for the hyperparameters to approximate optimal inputs well which, in turn, can be expected to affect feasibility and control performance. In the first approach, the shape of the sets is completely flexible and the "resolution" of this flexibility is limited only by the chosen grid (or more generally set) of training data. If the second approach is applied in its basic form, some rather large sets $\hat{\mathcal{D}}_i$ result if the desired feasible set $\mathcal{X}_f$ has large facets and the shape of the sets $\hat{\mathcal{D}}_i$ is not completely free. The sets $\mathcal{X}_f$ can have large facets first of all if its shape is restricted by state constraints (see Example 4.17). Clearly, the approach could be adopted such that large simplices are subdivided. In the third approach, the sets $\hat{\mathcal{D}}_i$ can be chosen completely freely within $\hat{\mathcal{X}}_f$. Yet, as each p_i has to be feasible for all $x \in \hat{\mathcal{X}}_f$, in fact this means that there is no freedom in terms of achieving feasibility. The flexibility in the third approach is further restricted as all parts p_i share a common matrix $\mathcal{K}_i$. Summarizing these findings, the first approach provides the largest flexibility with respect to the parametrization obtained, closely followed by the second approach.

All discussed comparisons are summarized in Table 4.4. As a conclusion, the second approach provides advantages similar to the first approach, yet, generally it is computationally considerably less demanding. Thus, whenever it is applicable (typically about $n \leq 6$ states), it should be the first choice. Only if it is not applicable, the third approach is to be preferred.

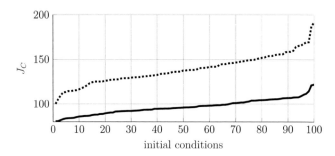

Figure 4.10: Comparison of the sorted achieved objective values J_C in the clustering using constant part (dotted line) and not using constant part (solid line) over different clustering initial conditions.

4.5.3 The effect of the constant term in the parametrization

Before we turn to the general evaluation of the semi-explicit MPC approach for linear systems, we evaluate the effect of computing parametrizations with constant part, i.e., allowing for $a_i \neq 0$. To this end, we reconsider Example 4.17.

Example 4.30 (Example 4.17 revisited). *We applied the first approach to compute feasible parametrizations (enlarging clusters and refinement) exactly as described above but with the additional degrees of freedom of allowing for $a_i \neq 0$. Interestingly, even though the parametrizations in this case are more general and include the previous case $a_i = 0$, not only did the results not improve, but, generally they even became worse. No feasible parametrization using a scalar decision variable $q = 1$ could be found. For the cases $q = 2$, $K \in \{2, 3\}$ automatically parametrizations with $a_i = 0$ were found (and could be rendered feasible), for $q = 2$, $K = 4$ a parametrization with $a_i \neq 0$ was found and could be rendered feasible.*

Tracing these issues back, also the objective J_C in the clustering algorithm showed deteriorated behavior. Figure 4.10 illustrates the sorted achieved value of J_C over 100 clustering initial conditions setting $a_i = 0$ (solid line) and allowing $a_i \neq 0$ (dotted line), respectively, in both cases using $q = 1$, $K = 5$. Surprisingly, approximately 70 % of the initial conditions using $a_i = 0$ yielded a lower objective value than the best objective value achieved allowing for $a_i \neq 0$. Beyond that, the best and the worst value achieved differ considerably more in the case of using $a_i \neq 0$. Both findings indicate that the constant part affects convergence of the clustering algorithm negatively. This hypothesis is supported by the observation that using $a_i \neq 0$ in many cases non-symmetric clustering patterns in the state space are observed even though one would expect to find the completely symmetric problem reflected in a symmetric clustering pattern of the states.

Note, furthermore, that Figure 4.10 hints that the clustering algorithm does not find a globally optimal solution for this problem even in the case $a_i = 0$ as the curve of achieved J_C values is not flattened out at the bottom end.

The results reported for the latter example are representative for the case of using $a_i \neq 0$ in the sense that qualitatively similar observations were made for most other examples considered. An intuitive explanation for the bad convergence behavior using a constant parts is that this constant part in the parametrization is "highly specific" for a certain set of states as it does not depend on the states. Once this constant has been chosen, a part p_i of the parametrization does not fit for most of the states and is thus not affected by them during the clustering procedure. This simply means that the algorithm is stuck in a (generally largely suboptimal) local minimum. Beyond that, intuitively it is clear that strongly tailoring a part p_i for some states, more parts p_i are needed to cover all states well. This aspect becomes obvious considering that generally parametrizations with constant part cannot be applied in sets which are shaped symmetrically with respect to the origin which is in contrast to parametrizations without constant part.

Some steps to improve convergence using constant parts in the parametrization could be taken beyond simply running the clustering algorithm from more initial conditions. For example, any additional knowledge available about the data to be clustered could be used. In particular, symmetric shapes in the clustering patterns could be fostered either within a cluster if $a_i = 0$ has appeared or via having two clusters located symmetrically to each other if $a_i \neq 0$. Furthermore, the constant part could be generally "weighted lower" in some way or its weight could be slowly increased during clustering. Some of these measures were implemented and tested and showed to actually improve convergence of the clustering to some extent. As the benefit of using constant parts still remained limited and it required a complicated procedure, this thread was not followed any further and in the linear case mostly parametrizations of the form $p_i(x, \bar{U}) = M_i \bar{U} + \mathcal{K}_i x$ were computed and used.

4.5.4 General evaluation of semi-explicit MPC for linear systems

Let us start our general evaluation of the semi-explicit MPC scheme for linear systems with the observation that this approach actually satisfies what has been suggested in (Alessio and Bemporad, 2009): " [...] semi-explicit methods should be also sought, in order to pre-process of line as much as possible of the MPC optimization problem without characterizing all possible optimization outcomes, but rather leaving some optimization operations on-line." In all numerical examples considered in this chapter, the semi-explicit approach actually enabled to offline prepare the online optimization by determining suitable parametrizations. The method was applicable to problems where computing and using the explicit solution is not possible or not desirable from an application point of view. Thus, the method is not only a conceptually interesting step forward which follows the semi-explicit idea but it actually contributes to the collection of available fast MPC schemes.

It is clear that any approach which addresses preparation of the arising optimization problem in a global manner has limited scalability in the problem size. In the presented numerical examples it became apparent that the desire to establish feasibility guarantees for the parametrized optimization even when state and input constraints are present is the main limiting factor for scalability. Main factors influencing scalability of the approach were the chosen method to compute feasible parametrizations and the complexity of the desired feasible set. Scalability of the method to problems where computing and using the exact explicit solution of the original mpQP is not reasonable or not possible was shown. Omitting feasibility guarantees (as is done for typical move-blocking strategies) would simplify the whole approach considerably and extend its scalability.

A key ingredient to theoretically prove stability of the origin of the closed loop was the candidate solution of the optimization problem obtained by shifting the previous solutions and appending the terminal control law. Yet, in the numerical examples considered, it turned out that none of the parametrizations which achieved high control performance required the candidate solution at all. Both the optimization remained feasible and open-loop costs of newly computed sequences automatically were sufficiently low to achieve a decrease in the open-loop costs along closed-loop trajectories. This hints that in a practical application the scheme could be simplified by not considering candidate solutions explicitly without loosing closed-loop asymptotic stability. Furthermore, in numerical tests parameters beyond τ were identified which promote the usage of newly optimized solutions. The most intuitive approach was to weight states closer to the origin higher in the clustering procedure. This results in a better approximation of optimal solutions for these states. Thus, state trajectories during closed-loop simulations gradually move into an area where the optimization benefits from the improved approximation. Correspondingly, chances increase that the new solution is better than the continued old solution and, as a result, the candidate solution is used less frequently.

The most important parameters in the semi-explicit approach clearly are the values of the hyperparameters K and q. As was observed in the numerical examples, the effect of these parameters is rather clear: Increasing either of them generally improves feasibility of the refinement when computing the parametrization and it improves control performance of the resulting semi-explicit MPC scheme. This correlation is rather strong, nevertheless, it is non-monotonic. This might be due to several reasons. First, the clustering algorithm disregards constraints of the mpP to be parametrized and, thus, the refinement step can deteriorate approximation accuracy of optimal input sequences. Second, open-loop costs using parametrizations translate into closed-loop performance in a nontrivial way, see also the discussion on the general semi-explicit MPC approach in Subsection 3.5.3. Third, using high values for K and q but insufficiently rich training data will lead to over-fitting of the training data. This results in high approximation accuracy of the training data, but, bad generalization to new states.

During the work on this thesis, it turned out that a relevant characteristic of the semi-explicit MPC approach is the richness in adjustment possibilities and degrees of freedom it possesses in the whole offline procedure, also beyond K and q. These parameters comprise among others the training states used (amount, random selection/selection from grid (rotated/aligned) (with/without perturbation added)), weighting of training states, horizon length, weighting of input sequence along horizon, desired feasible set, feasibility guaranteeing procedure 1/2/3, weighting in the refining optimization, minor parameters affecting numerics in the whole work flow, etc.. Whereas the effect of some of the parameters is rather clear and predictable, for some of them the effect is not completely clear rendering their selection less intuitive. On the one hand having too many degrees of freedom might make application of the method more difficult. On the other hand, the degrees of freedom are a manifestation of the large flexibility and customizability of the general method and, even limiting the parameters to the most important ones, the general approach is still effective and powerful.

4.5.5 Comparison of semi-explicit MPC to existing simplified MPC schemes

Next, we compare the semi-explicit MPC approach to existing efficient MPC schemes and then evaluate advantages and disadvantages of the method. In practical applications typically MPC schemes are employed which have been generically simplified over the ones considered in theoretical contexts by omitting some of the stability guaranteeing ingredients. Mainly this means that no terminal constraints and rather short prediction horizons are used. This results in the important difference with respect to the semi-explicit approach as proposed here that these schemes do not attempt to establish any theoretical feasibility or closed-loop stability guarantees. As not establishing such guarantees would simplify the semi-explicit MPC approach vastly, the following comparison will mainly comprise schemes which are equipped with such guarantees. In spite of that, as already mentioned above, a semi-explicit scheme could be applied in such simplified setting as well.

Comparison to other parametrizations

Recall that applying parametrizations in an MPC scheme, the goal is typically to improve the tradeoff between the region of attraction, control performance and online computational burden (Khan et al., 2014). Most available results address these objectives and differ in how they are weighted against each other and in whether or not guarantees on (recursive) feasibility and asymptotic stability are given.

The simplest case of a parametrization is a move-blocking strategy which fixes online complexity via the number of decision variables used but does not explicitly address control performance, feasible set and closed-loop stability. Investing some offline computations, other approaches allow to maximize the feasible set for given complexity (e.g. Shekhar and Manzie (2015)) or essentially fix a feasible set and aim at maximizing control performance (e.g. Li et al. (2013)). Based on a more complex offline optimization, an approach to find parametrizations approximately at the Pareto frontier between a measure for control performance, the feasible set and the dimension of the decision variable is presented in (Khan et al., 2014).

In the proposed semi-explicit MPC approach we fix a desired feasible set and then take the complexity of the parametrizations as tuning knob. In addition to the dimension of the decision variable, complexity comprises the number of parts of the parametrization expressed via K. Control performance is addressed implicitly when the preliminary parametrizations are computed and explicitly in the refinement scheme and it can be increased by increasing complexity of the parametrizations as was seen in the examples. The refinement scheme is comparable to the optimization used in (Khan et al., 2014).

Most existing simplifying parametrizations do not depend on the system state they are applied for as generally and explicitly as the ones proposed here. An exception are interpolation based schemes which can be seen as state dependent-parametrizations. Different interpolation based schemes exists of which some take a desired feasible set as a starting point (e.g. (Rossiter and Grieder, 2005)) and others fix the number of degrees of freedom used in the interpolation (e.g. (Rossiter et al., 2004)). The schemes also differ in the optimization problem which has to be solved online, yet, in fact a common feature is that for the optimization some state-dependent information on (typically feasible) solutions is prepared offline and stored and exploited online. This feature and the range of applicability is a similarity of interpolation based schemes and the proposed method. As mentioned above, an interpolation based scheme is covered to some extent as a special case by the parametrizations presented here.

Rather few of the existing parametrizations guarantee feasibility of the original optimization problem for a given set of states when unaltered state, input and terminal constraints are present. This is achieved by the parametrizations presented here. Clearly, the explicit state-dependence and the feasibility guarantee for a given set are achieved via more expensive offline computations than required by most of the existing methods. Nevertheless, for the considered examples offline computations were still well tractable as the given computation times show.

For the proposed semi-explicit approach, typically decision variables of considerably lower dimension than required in other parametrization based approaches are possible. The discussion on the complexity of numerically solving a parametrized optimization problem highlighted the benefits thereof. Many existing results suggest to include the candidate input sequence as additional basis vector in the parametrization (Li et al., 2013; Shekhar and Manzie, 2015; Valencia-Palomo and Rossiter, 2011). This requires at least a two-dimensional decision variable. Other schemes as for example (Li et al., 2013) and move-blocking schemes require at least one decision variable (and typically even several ones) per system input. In the semi-explicit approach, the possibility to use a piecewiese defined parametrization and to make it state dependent allows to use scalar decision variables even for systems with more than one scalar input.

Comparison to suboptimal explicit MPC results

As observed in several of the examples, the proposed method is applicable to problems where the exact explicit solution of the multi-parametric optimization problem is highly and possibly prohibitively complex. In such cases, applying a suboptimal explicit solution can also be a remedy. Suboptimal explicit approaches can be subdivided into two main categories: Approaches which require the exact explicit solution in order to compute a simplification thereof, for example (Kvasnica et al., 2011, 2013), and approaches which work directly on the problem data as for example (Jones and Morari, 2009, 2010; Summers et al., 2011). In terms of maximum scalability of the methods, the latter approaches are superior. Nevertheless, all suboptimal explicit approaches have in common with the semi-explicit approach that they require a relevant amount of offline computations which become increasingly expensive with growing state space dimension.

In comparison to the proposed method, suboptimal explicit approaches seem to provide more flexibility in the shape of the feasible set and their complexity seems to behave better for complex desired feasible sets. This might be beneficial first of all for higher dimensional systems. The proposed method, on the other hand, seems to have advantages in terms of minimum required "quality" of the solutions. Whereas suboptimal explicit schemes typically require that each affine feedback law achieves a decrease in $J^*(x)$ to ensure closed-loop asymptotic stability, for the proposed method it suffices to provide a feasible solution for each state as system theoretic properties can be exploited to extend the feasible solution into a stabilizing one. This hypothesis is also supported by the observed numbers of critical regions in Example 4.28 and the comparison which is drawn to the referenced results. So the semi-explicit approach might be beneficial if one is interested in schemes of minimal complexity such that still closed-loop stability guarantees can be established.

Suboptimal explicit schemes have the advantage of extremely fast and totally predictable online computation times. For the proposed method, this can be partly achieved if the almost explicit scheme based on scalar decision variables presented in the next chapter or the explicit solutions of the parametrized optimization problems are applied.

4.5.6 Strengths and drawbacks of the semi-explicit MPC approach for linear systems

Next, we first summarize properties in which the semi-explicit MPC is inferior to other methods before we turn to its advantages.

- Even though simplifying the online procedure in the sense that lower-dimensional optimization variables can be used, the procedure is conceptually complicated over approaches which employ one globally defined parametrization as not only a numerical optimization has to be executed but also further operations and decisions have to be made.

- The offline part of the proposed approach has limited scalability and can become inapplicable for systems with too high-dimensional states. Yet, scalability of the proposed method has to be related to the scalability of comparable approaches.

- Maximum achievable control performance via the semi-explicit approach is limited, first of all for larger problems. This is due to the fact that only limited information about optimal solutions can be sampled and that there is no definite connection of open-loop and closed-loop performance. This is in contrast to, for example, simplified explicit solutions which typically can be chosen to approximate the optimal solution up to any arbitrary accuracy.

- The chosen offline approach requires to start out working globally on the complete problem data. This fact is a structural limitation of the proposed method which restricts the total amount of data which can be used in the proposed offline procedure. As a consequence, both previously named issues arise, i.e. limited scalability and limited maximum achievable control performance. Suboptimal explicit MPC schemes which work hierarchically or iteratively behave better in this respect as a successive refinement and inclusion of more data is possible.

- Applying the method, the outcome is affected by many parameters which can be adjusted by the user. This can make employing the approach non-straightforward.

The following are advantages and features of the semi-explicit MPC approach which make it superior to other approaches.

+ Guarantees for feasibility and closed-loop asymptotic stability are provided. This is in contrast to many simpler parametrizations and ad-hoc simplified MPC schemes.

+ A pre-specified set of states can be included into the feasible set of the parametrized optimization and into the region of attraction of the corresponding closed loop. As for the previous item, this is the case for only few of the parametrized MPC schemes.

+ For a given MPC problem and desired region of attraction typically the complexity of the simplest solution obtained applying the semi-explicit MPC approach is very low. This holds true independently of the chosen representation, i.e., it either results in a very low-dimensional decision variable or in a low number of critical regions of the corresponding explicit solution. This advantage is achieved by storing state-dependent information on the solutions (in comparison to other parametrizations) with the system theoretic extension of the solutions into a stabilizing scheme (in contrast to simplified explicit solutions which typically are required to be directly stabilizing).

+ The semi-explicit approach scales better than optimal explicit approaches both regarding the offline procedure and regarding the online applicability of the solution. This is made possible by computing suboptimal solutions and by the combined offline preparation and online optimization procedure. In particular, the method is superior to explicit approaches for problems with high-dimensional decision variable resulting from long prediction horizons and/or high-dimensional system inputs.

+ The semi-explicit approach works on a structural level which makes it beneficial rather independently of the problem details and the chosen online solution strategy.

4.6 Summary and extensions

4.6.1 Summary

In this chapter, we presented further and more detailed results on the semi-explicit MPC approach for linear systems. In particular, we presented three approaches to compute parametrizations for a given MPC problem which are guaranteed to result in a feasible parametrized optimization problem for a given set of states. The approaches have different strengths is terms of scalability, offline complexity and capability of the resulting parametrization to approximate optimal solutions of an mpP. This was evaluated theoretically and illustrated in a numerical example. We established an upper bound on the relative closed-loop cost increase incurred due to application of the semi-explicit MPC scheme. An offline stability check for the parametrization was presented which can verify closed-loop asymptotic stability of a simplified online semi-explicit MPC scheme. The whole semi-explicit MPC approach for linear systems was evaluated theoretically and compared to existing approaches in detail. Several numerical examples were presented to illustrate application and characteristics of the results. Thereby, the effects of the most important tuning knobs of the method were evaluated and advantages of the method over existing approaches in terms of scalability and flexibility in the design were shown.

4.6.2 Extensions

The semi-explicit MPC scheme as proposed in this thesis and elaborated for linear systems in the current chapter constitutes a conceptually new approach to the goal of simplifying MPC schemes. Accordingly, it provides numerous starting points for extensions, further investigations, results and fields of application. In the following, some possible direct extensions of the approach for linear systems are given.

Further approaches to find feasible parametrizations

Clearly, extensions and refinements of the approaches to find feasible parametrizations are possible and could further extend applicability of the general method.

The first approach to compute feasible parametrizations based on enlarging convex hulls of clustered states by hypercubes is structurally simple and amenable to various adaptations. The approach would simplify drastically if the hypercubes were not intersected with the desired feasible set but hypercubes which are not completely contained in the desired feasible set were just discarded. Not mapping the clustered states to the corners of a hypercube for the refinement but working with the clustered training data directly would simplify the refining optimization. Instead, techniques from robust MPC as pre-stabilizing the system and pre-tightening of the constraints could be used in order to nevertheless ensure feasibility on the hypercubes around the states. This approach is similar to the one proposed for nonlinear systems in Chapter 6 of this thesis. Another adaptation of the general approach is to refine the grid of training data used locally at the edges of the clusters in order to refine the shape of the clusters in state space. Investigations in this direction have been done in (Alber, 2013).

The second approach to find feasible parametrizations based on clustering simplices at the facets could directly be extended to refining the simplices by subdividing large ones or ones which complicate achieving feasibility. Also the contrary is possible, namely, to combine the vertices of several simplices and to tread them together in order to reduce the computational effort. More generally, clustering objects other than simplices both at the edge of the desired feasible set or volumes from the interior of the desired feasible set are possible.

The third approach to compute feasible parametrizations based on finding an independently feasible part solving a linear program requires in many cases the application of prediction horizons which are long with respect to the feasible set which is achieved eventually. Using at the same time a simple parametrization, this can deteriorate control performance for states closer to the origin. An extension remedying this issue could consist in using two types of parametrizations in parallel: One based on a long prediction horizon yielding a large feasible set and a second one based on a shorter prediction horizon which contributes control performance for states closer to the origin. Both parts could be seamlessly used in parallel and the decision for one of the parts would be made automatically based on the observed open-loop control performance. Furthermore, parsimonious selection of the states used in the linear program contributes to scalability of this approach. Further investigations could address how to ensure that the obtained feasible set is large using as few states as possible.

Exploiting symmetries in the problem

For explicit MPC some results are available on exploiting symmetries in the problem in order to simplify matters (Danielson and Borrelli, 2015). The semi-explicit approach could benefit from corresponding approaches in two ways. On the one hand, symmetries could enable to omit some of the training data in the offline procedure if the information this data carries is also contained in a subset of the data. Clearly, this would simplify the offline procedure and improve its scalability. Similarly, symmetries in the problem data could result in symmetric patterns in the parametrizations computed which would make simplification thereof possible.

Ensuring for the second approach to compute feasible parametrizations that the simplices at the facets to be clustered are located symmetrically with respect to the origin was a first step towards this goal. Another step in this direction which has been already taken was in the third approach to find feasible parametrizations to avoid using symmetrically located states in the linear program to be solved but to remove redundant data in such cases.

Application of the approach to reduce the number of constraints

So far, the parametrizations were applied aiming mainly at reducing the dimension of the decision variable to a desired value. Applications could also be thought of where the main benefit is generated by the resulting reduction in the number of constraints in the mpP. For example if the optimization in min-max MPC is formulated as a quadratic program (de La Pena et al., 2007), a large number of constraints results. It would be interesting to see if applying the proposed parametrizations can achieve a significant reduction in the number of non-redundant constraints. Beyond that, it might be worthwhile to examine more in detail how the employed parametrizations affect the resulting number of non-redundant constraints, so that parametrizations could be sought which achieve a particularly large reduction.

Application of the linear methods to more general problem classes

It is insightful to identify the core ingredients which enabled to apply the methods of this chapter to linear systems. In fact, most of the results found primarily on joint convexity of the constraints (and objectives) in the state and the vector of stacked predicted inputs. Clearly, considering linear system dynamics and polytopic constraints is sufficient for this property, yet, by no means this is necessary. As is done in some results on explicit MPC for nonlinear systems (e.g. Johansen (2004)), the scope could easily be broadened to general nonlinear systems which result in multi-parametric optimization problems which have jointly convex constraints.

Beyond the mentioned extensions of the approach itself, it would be worth to examine more in detail how the parametrized optimization problems can be solved efficiently. The most interesting and relevant special case clearly is obtained if parametrizations are used which possess only a scalar parameter resulting in optimization problems with scalar decision variable. The next chapter is dedicated to this topic.

Chapter 5

Almost explicit MPC for linear systems

The semi-explicit MPC approach presented in this thesis and elaborated for linear systems in the previous chapter allows to drastically reduce the dimension of the decision variable in the online optimization problem in MPC. In many cases even a scalar decision variable is sufficient. So far the benefit of these results was based on the general simplification of numerical optimization procedures that comes with decision variables of reduced dimension. Our goal in the current chapter is to go beyond such gradual simplification by presenting an extremely simple numerical optimization strategy which exploits the specific simplicity of having a *scalar decision variable*, i.e, $q = 1$. In order to achieve this simplification, the properties of this type of problem will be exploited and further computations will be shifted offline to prepare the problem for its online solution. As a result of further reducing online computational load, application of the semi-explicit MPC scheme is rendered even more attractive. The results of this chapter are applicable to the mpQPs resulting from linear systems with quadratic stage and terminal cost functions.

This chapter is partly based on (Goebel and Allgöwer, 2015, 2017a).

5.1 Introduction and problem formulation

In this chapter we consider a convex multi-parametric quadratic program in a scalar decision variable given by

$$\min_{\tilde{u} \in \mathbb{R}} \tilde{H}\tilde{u}^2 + x^\top \tilde{F}\tilde{u} + x^\top \tilde{Y} x$$
$$\text{s.t.} \quad \tilde{G}\tilde{u} \leq W + \tilde{E}x \tag{5.1}$$

with $x \in \mathbb{R}^n$. We assume $\tilde{H} > 0$ and will comment in the end on the case $\tilde{H} = 0$ which in fact reduces the objective function to an affine one. The latter general mpQP is linked to the results presented in the previous chapter assuming a parametrization of the form[1] $p_i(x, \tilde{u}) = M_i\tilde{u} + \mathcal{K}_i x$ via the identities $\tilde{H} = M_i^\top H M_i$, $\tilde{F} = 2\mathcal{K}_i^\top H M_i + F M_i$, $\tilde{Y} = Y + F\mathcal{K}_i + \mathcal{K}_i^\top H \mathcal{K}_i$, $\tilde{G} = G M_i$ and $\tilde{E} = E - G\mathcal{K}_i$, see e.g. (4.6). Note that in the following the indices i are omitted and the results apply to each part p_i of a parametrization likewise. If (5.1) has been obtained in a semi-explicit MPC scheme via these relations, the

---

[1] The case which arises from using a parametrization with constant term (i.e. $a_i = 0$) is not considered here explicitly as it would complicate matters while not providing any additional insight and being of limited practical relevance. Corresponding extensions are straight forward, for example via using an extended state $\tilde{x} = [x^\top, 1]^\top$ and a suitably transformed matrix $\mathcal{K}_i$.

assumption $\tilde{H} > 0$ is generally fulfilled as H is positive definite (see e.g. (Tøndel et al., 2003)) and M_i is not a zero vector. As in previous chapters, having the semi-explicit MPC scheme in mind, we keep calling x state, being aware that in a semi-explicit MPC scheme x might comprise quantities beyond states and that the results also remain valid beyond the application in semi-explicit MPC schemes where x is a general parameter. As the solution strategy presented in this chapter shifts further computational load offline and simplifies online operations further, we call the method *almost explicit* solution strategy. Correspondingly, a semi-explicit MPC scheme which applies the almost explicit solution strategy will be called *almost explicit MPC*.

In this chapter, we first present the basic almost explicit solution strategy, then introduce two simplifications thereof and finally evaluate its numerical complexity theoretically as well as in a numerical example and compare its complexity to the complexity of two different explicit solution strategies.

5.2 Basic almost explicit strategy

Before we turn to the basic almost explicit MPC approach in detail, let us look at a result which captures the general principle behind the results in this chapter.

Proposition 5.1. *Consider the optimization*

$$
\begin{aligned}
J^* = \min_{z \in \mathbb{R}} &\ J(z) \\
s.t. &\ z \in [\underline{z}, \overline{z}]
\end{aligned}
\tag{5.2}
$$

with convex objective function $J : \mathbb{R} \to \mathbb{R}$ and $\underline{z} \leq \overline{z}$ defining the constraint set. Let $\tilde{z} \in \mathbb{R}$ be such that $J(\tilde{z}) \leq J(z)$ for all $z \in [\underline{z}, \overline{z}]$. For

$$
\begin{aligned}
z^* = \arg\min_{z \in \mathbb{R}} &\ |\tilde{z} - z| \\
s.t. &\ z \in [\underline{z}, \overline{z}]
\end{aligned}
\tag{5.3}
$$

it holds that $J^* = J(z^*)$. In particular, at least one element of $\{\underline{z}, \overline{z}, \tilde{z}\}$ is a minimizer of (5.2).

Proof. Due to optimality $J(z^*) \geq J^*$ holds. Thus, we show $J(z^*) \leq J^*$ to conclude that $J^* = J(z^*)$ as claimed. Let v be a minimizer of (5.2). If $\tilde{z} \leq v$, it holds due to (5.3) that $\tilde{z} \leq z^* \leq v$ (and correspondingly $v \leq z^* \leq \tilde{z}$ if $v \leq \tilde{z}$). Let $z^* = \alpha \tilde{z} + (1 - \alpha)v$ for suitable $\alpha \in [0, 1]$. We have the relation

$$
J(z^*) = J(\alpha \tilde{z} + (1 - \alpha)v) \leq \alpha J(\tilde{z}) + (1 - \alpha)J(v) \leq J(v) = J^*
$$

where the first inequality is due to convexity of J and the second inequality is due to the definition of $\tilde{z}$ and v. $\qquad\square$

In the proposition, $\tilde{z}$ can be a global minimizer of J (if it is finite) or any real number which yields a lower bound $J(\tilde{z})$ on J^*. The benefit of this proposition is as follows: It states that a simple case analysis comparing three numbers is sufficient to solve a convex optimization problem in a scalar decision variable for which a value of the decision variable is known which bounds the achievable objective from below. Clearly, for linear and quadratic objective functions such value is readily available. In comparison to applying a generic numerical optimization algorithm, exploiting the latter result simplifies the numerical solution of such problems drastically. The remainder of this chapter is dedicated to an in depth investigation of the parameter-dependent quadratic version of this problem as it arises from linear MPC problems with quadratic objective function and affine constraints.

Let us now turn to this quadratic case and consider the scalar mpQP (5.1). Completing the squares via a coordinate transformation $\tilde{u} = z - \frac{1}{2}\tilde{H}^{-1}\tilde{F}^\top x$, an equivalent optimization

$$\min_z \tilde{H}z^2 + x^\top \overline{Y}x \\ \text{s.t. } \tilde{G}z \leq \overline{E}x + W \tag{5.4}$$

with purely quadratic objective can be formulated in the new decision variable z wherein $\overline{Y} = \tilde{Y} - \frac{1}{4}\tilde{H}^{-1}\tilde{F}\tilde{F}^\top$ and $\overline{E} = \tilde{E} + \frac{1}{2}GM\tilde{H}^{-1}\tilde{F}^\top$ holds. Note that $\tilde{H}^{-1}$ exists as $\tilde{H} > 0$ was assumed. The goal of this transformation is to ensure that the (global) optimizer of the corresponding unconstrained problem is at $\tilde{z} = 0$ independently of x, c.f. Proposition 5.1.

Next, the constraints in (5.4) are reorganized. Elements in $\tilde{G}$ can be positive, negative or zero, corresponding to upper bounds on z, lower bounds on z and constraints on the state x alone, respectively. Sorting the constraints accordingly and normalizing them in the first two cases row-wise by the entries in $\tilde{G}$, the following equivalence results

$$\tilde{G}z \leq \overline{E}x + W \quad \Leftrightarrow \quad E^l x + W^l \leq \mathbb{1}z \wedge \mathbb{1}z \leq E^u x + W^u \wedge E^x x \leq W^x \tag{5.5}$$

for suitable matrices W^l, E^l, W^u, E^u, W^x and E^x. Using this equivalence, for x such that $E^x x \leq W^x$ holds, the optimization

$$\min_z z^2 \\ \text{s.t. } \max(E^l x + W^l) \leq z \leq \min(E^u x + W^u) \tag{5.6}$$

yields the same optimizer as QP (5.4) (therein, $\max(\cdot)$ and $\min(\cdot)$ in the constraints refers to the maximum and the minimum entry of the vector it is applied to, respectively). In particular, this means that QP (5.4) can be solved by considering the interval of feasible z-values and, if this interval is non-empty, choosing the element therein with lowest absolute value. As this is either at the boundaries or at zero, it suffices to consider the numbers $b_l = \max(E^l x + W^l)$, $b_u = \min(E^u x + W^u)$ and 0. This is formalized in Algorithm 13.

The solution of (5.1) can be recovered from the output of Algorithm 13 according to the following proposition.

Proposition 5.2. *Algorithm 13 yields a solution z^* if and only if mpQP (5.1) is feasible. If (5.1) is feasible with solution $J^*(x)$ and optimizer $\tilde{u}^*$, it holds that*

$$J^*(x) = \tilde{H}(z^*)^2 + x^\top \overline{Y}x$$

and $\tilde{u}^* = z^* - \frac{1}{2}\tilde{H}^{-1}\tilde{F}^\top x$.

Summarizing, in order to solve (5.1) for a given state x, no numerical optimization procedure in the classical sense is required but a simple case analysis as stated in Algorithm 13 is sufficient. Most of the quantities involved therein can be pre-computed simplifying the actual numerical optimization drastically. A detailed complexity evaluation of Algorithm 13 will be given below after further simplifications have been introduced.

Algorithm 13 Almost explicit solution of QP (5.4)

Input: W^l, E^l, W^u, E^u, E^x, W^x, state x
  1: set $b_l = \max(E^l x + W^l)$
  2: set $b_u = \min(E^u x + W^u)$
  3: **if** $b_l \leq b_u \wedge E^x x \leq W^x$ **then** \\QP (5.4) feasible
  4:     **if** $0 < b_l$ **then**
  5:         $z^* = b_l$
  6:     **else**
  7:         **if** $0 < b_u$ **then**
  8:             $z^* = 0$
  9:         **else**
10:             $z^* = b_u$
11:         **end if**
12:     **end if**
13: **end if**
Output: z^* if existent

Remark 5.3. *Applying the almost explicit solution strategy in a semi-explicit MPC context, one is eventually not interested in $\tilde{u}^*$ but in evaluating the parametrization $p(x, \tilde{u}^*) = M\tilde{u}^* + \mathcal{K}x$. In this case it is reasonable to skip the intermediate $\tilde{u}^*$ computation and directly evaluate*

$$p_i(x, \tilde{u}^*(\tilde{z}^*)) = Mz^* + \tilde{\mathcal{K}}x$$

where $\tilde{\mathcal{K}} = \mathcal{K} - \frac{1}{2}M\tilde{H}^{-1}\tilde{F}^{\top}$ holds. Thus, employing the mpQP in z coordinates does in fact not introduce any additional computational overhead for the online semi-explicit MPC scheme. This way Algorithm 13 can be employed to obtain an admissible and optimized predicted input sequence and to evaluate the corresponding open-loop cost.

5.3 Simplified almost explicit strategy

The almost explicit solution algorithm can be simplified further if non-required constraints are omitted or if non-required computations are avoided. These tasks are addressed next.

5.3.1 Removing non-required constraints

The main computational burden of Algorithm 13 is the computation of the lower and upper bounds on z in lines 1 and 2. Hence, in order to reduce the computational load caused by Algorithm 13, reducing the number of constraints is the most effective measure to take. To this end, we first have a closer look at the different types of constraints in (5.4) and then identify non-required ones to be removed. In the following, we use the projection $\mathcal{P}_x = \{x | \exists z : \tilde{G}z \leq \overline{E}x + W\}$ which is the set of states x for which (5.4) is feasible. Let its hyperplane representation be given by $\mathcal{P}_x = \{x | \tilde{E}^x x \leq \tilde{W}^x\}$. Such projection can be found for example via application of the Fourier-Motzkin elimination (Keerthi and Sridharan, 1990). Four types of constraints in (5.4) can be classified:

- *Type a) constraint*: A constraint which defines the optimal z^* for at least one x, i.e.,

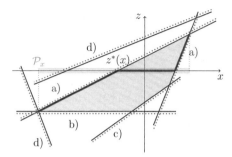

Figure 5.1: Illustration of constraint types in QP (5.4). The gray area corresponds to the set $\{(x,z)|\tilde{G}z \leq \overline{E}x + W\}$.

the constraint is active for this x and removing the constraint would change $z^*(x)$, is called type a) constraint.

- *Type b) constraint*: A constraint which is not of type a) and constrains the set $\mathcal{P}_x$ is called type b) constraint.

- *Type c) constraint*: A constraint which is neither of type a) nor of type b) and defines the set $\{(x,z)|\tilde{G}z \leq \overline{E}x + W\}$ is called type c) constraint.

- *Type d) constraint*: All redundant constraints are called type d) constraint.

Figure 5.1 illustrates the types of constraints for a fictitious example with $x \in \mathbb{R}$. Note that type a) constraints are those constraints for which an x exists such that they are strongly active at $x, z^*(x)$. Before we address how to identify the types of constraints, we state a result which makes use of the classification of the constraints.

Proposition 5.4. *Let a and b denote the set of all indices of type a) and type b) constraints of QP (5.4), respectively. Consider the following two QPs:*

$$QP1: \qquad \min_z \tilde{H}z^2 + x^\top \overline{Y}x$$
$$\text{s.t. } \tilde{G}_j z \leq \overline{E}_j x + W_j, \quad j \in a \cup b$$

$$QP2: \qquad \min_z \tilde{H}z^2 + x^\top \overline{Y}x$$
$$\text{s.t. } \tilde{G}_j z \leq \overline{E}_j x + W_j, \quad j \in a$$

$$\tilde{E}^x x \leq \tilde{W}^x.$$

QP1 and QP2 are equivalent to QP (5.4) in the sense that they are feasible for the same set of x and for feasible x, they share the same optimizer and the same optimal value.

Proof. Equality of the feasible sets follows from the definition of the constraint types and of the matrices $\tilde{E}^x$, $\tilde{W}^x$. By definition, only removal of type a) constraints changes the optimizer. As all QPs share the same type a) constraints, they share the same optimizer and, hence, the same optimal value. $\qquad \square$

Note that the equivalence does not hold in terms of admissible z values for a given x. This might be an issue if the QPs are solved via a numerical optimization algorithm which is not guaranteed to reach the optimum but which might terminate prematurely at a feasible solution which in fact might be infeasible with respect to the original problem. Yet, Algorithm 13 (and also Algorithm 14 below) attains the optimum and in fact benefits from the equivalent reformulations. Due to the definition of the constraint types and of the constraints used in QP1, this formulation comprises the minimum amount of constraints. Therefore its numerical solution is simpler and storage requirements are reduced with respect to QP (5.4). Typically, QP2 requires more constraints than QP1, yet, QP2 will be used for a comparison with an explicit solution of QP (5.4) and will be employed in a simplified scheme below.

Let us next address identification of the types of the constraints. As is well known, type d) constraints can be identified via solving one linear program per constraint and can be removed at no losses, see e.g. (Borrelli et al., 2015). Thus, in the following we assume that all type d) constraints have been removed. Type a) and type b) constraints can then be identified based on the following lemmas.

Lemma 5.5. *The following is equivalent:*

i) Constraint j is of type a).

ii) There is a pair (x, z) such that $\tilde{G}z \leq \overline{E}x + W$, constraint j is the only active lower (upper) bound at (x, z) and $z > 0$ ($z < 0$).

Proof. "¬ii) ⇒ ¬i)" Assume there is no pair (x, z) such that constraint j is the only active lower (upper) bound and $z > 0$ ($z < 0$). Then, constraint j can either not be active at an optimum or removing it would not change the optimum due to a second active lower (upper) bound or because $z^* = 0$. "ii) ⇒ i)" If such pair (x, z) exists, then z is the optimal solution for this x as z^2 cannot be decreased without violating constraint j. If constraint j is removed, z is no longer optimal for the chosen x as no other lower (upper) bound is active and z can be moved towards zero within the feasible interval. □

Note that due to continuity of the mappings defining the constraints, the latter result implies that each type a) constraint is active for a set of states with non-empty interior. Condition ii) given in the lemma can be checked via solving one linear program per constraint. In the next result, we call those constraints which define in (5.6) upper (lower) bounds on z upper (lower) bound constraints.

Lemma 5.6. *Assume constraint j is an upper (lower) bound constraint and not of type a). Constraint j is of type b) if and only if there exists a pair (x, z) such that $\tilde{G}z \leq \overline{E}x + W$ and constraint j is active at (x, z) together with a lower (upper) bound constraint r.*

Proof. A hyperplane representation of the set $\mathcal{P}_x$ can be obtained by Fourier-Motzkin elimination. Therein, each pairwise combination of upper bound j and lower bound r on z is merged into one constraint of $\mathcal{P}_x$ via $(E_r^l - E_j^u)x \leq W_j^u - W_r^l$. Such new constraint can become active if and only if both of the original constraints can become active at a common pair (x, z). Consequently, a constraint which can never become active together with an "opposite" bound can only yield redundant constraints on $\mathcal{P}_x$. □

The situation of the latter result is illustrated in Figure 5.1: The type c) constraint does not intersect any opposite bound at a feasible (x, z) combination, whereas the type b) constraint does so. Identifying type b) constraints via this criterion is numerically well tractable and requires less than n_t^2 linear programs to be solved where n_t is the total number of constraints in QP (5.4). Clearly, pure state constraints are either redundant or of type b).

Summarizing, in order to decide for a given state x whether (5.4) is feasible and if so to find the optimizer $z^*(x)$, only type a) and type b) constraints need to be considered and type c) and type d) constraints can be neglected. The foregoing results provide a way to a priori identify the types of the constraints so that type c) and type d) constraints can be removed from QP (5.4) when producing the input data for Algorithm 13. This way both memory requirements and computational requirements to numerically solve the optimization can be reduced and still the optimal solution of QP (5.4) can be recovered.

5.3.2 Avoiding non-required computations

We next consider another simplification of the almost explicit algorithm which is possible if the considered mpQP fulfills a special property formulated in the following assumption.

Assumption 5.1. *Consider mpQP (5.1). Assume that $\tilde{u}^f = 0$ is a feasible solution for all states x for which the problem is feasible.*

Whereas this assumption might seem rather restrictive at a first glance, it is typically fulfilled in the semi-explicit MPC context if a parametrization is used which has been computed via the third approach to find feasible parametrizations (Algorithm 10). This approach aims at ensuring that the obtained parametrization $p_i(x, \tilde{u}) = M_i \tilde{u} + \mathcal{K} x$ yields a feasible solution of the mpQP for a large set of states x when setting $\tilde{u} = 0$. So employing such parametrization and adding the constraint $x \in \hat{\mathcal{X}}_f = \{x | (G\mathcal{K} - E)x \leq W\}$ to the parametrized mpQP guarantees that Assumption 5.1 is fulfilled.

So let Assumption 5.1 be fulfilled. For feasible states x, a feasible solution of the parametrized mpQP (5.1) is given by $\tilde{u}^f = 0$ and correspondingly $z^f(x) = \tilde{H}^{-1} \tilde{F}^\top x$ is a feasible solution of (5.4). If $z^f > 0$, it follows that $z^* \geq 0$ and at z^* either no type a) constraint is active or a lower bound of type a) is active. The counterpart for $z^f < 0$ holds true correspondingly. Thus we have the following result.

Proposition 5.7. *Consider QP (5.4) and let Assumption 5.1 hold. Then, a hyperplane separating the set $\{x \in \mathcal{P}_x | z^*(x)$ is at a lower bound$\}$ from the set $\{x \in \mathcal{P}_x | z^*(x)$ is at an upper bound$\}$ is given by $\tilde{H}^{-1} \tilde{F}^\top x = 0$.*

Again working based on QP2, Algorithm 13 can be simplified exploiting this circumstance. This results in Algorithm 14. For this simplified almost explicit algorithm only either upper bounds or lower bounds have to be computed and evaluated for a given state x. A detailed complexity evaluation is given below.

Algorithm 14 Simplified almost explicit solution of QP (5.4)

Input: W^l, E^l, W^u, E^u, E^x, W^x, $\tilde{H}^{-1}\tilde{F}^\top$, state x
1: **if** $\tilde{E}^x x \leq \tilde{W}^x$ **then** \\QP (5.4) feasible
2:     **if** $\tilde{H}^{-1}\tilde{F}^\top x > 0$ **then**
3:         set $b_l = \max(E^l x + W^l)$
4:         $z^* = \max\{b_l, 0\}$
5:     **else**
6:         Set $b_u = \min(E^u x + W^u)$
7:         $z^* = \min\{b_u, 0\}$
8:     **end if**
9: **end if**
Output: z^* if existent

5.4 Examples and evaluation

5.4.1 Complexity evaluation and comparison to explicit solution

In the following, we first take a look at the numerical complexity of Algorithms 13 and 14. We then consider the explicit solution of QP (5.4) and establish a relation of the constraints of QP (5.4) and the complexity of its explicit solution. Finally, we compare the numerical effort of applying Algorithms 13 and 14 to the complexity of two algorithms based on the explicit solution. To simplify matters, we assume that representation QP2 is used and the feasibility check $\tilde{E}^x x \leq \tilde{W}^x$ is neglected for all compared algorithms.

Let N_l and N_u be the number of type a) upper and lower bounds in QP (5.4), respectively. Algorithm 13 requires computation of $n(N_l + N_u)$ multiplications, $n(N_l + N_u)$ sums and $N_l + N_u$ comparisons to find the max/min of the constraints and z^*. Algorithm 14 requires executing $n(\max\{N_l, N_u\} + 1)$ multiplications, $n(\max\{N_l, N_u\} + 1)$ summations and $\max\{N_l, N_u\} + 1$ comparisons. The corresponding summed numbers of floating point operations required are given in Table 5.1.

An alternative to applying Algorithm 13 or 14 to solve QP (5.4) would be to take the mpQP perspective, solve QP (5.4) explicitly a priori and then, for a given state x, evaluate this solution. Let the explicit solution of QP (5.4) be given as

$$z^*(x) = F_s x + f_s \text{ if } x \in \mathcal{CR}_s$$
$$\mathcal{CR}_s = \{x | A_s x \leq b_s\}, \quad s = 1, \ldots, n_{CR} \tag{5.7}$$

with $A_s \in \mathbb{R}^{c_s \times n}$ and $b_s \in \mathbb{R}^{c_s}$ defining full-dimensional critical regions. The following holds.

Lemma 5.8. *Consider QP (5.4) and its explicit solution (5.7). The following relations are valid.*

- *A critical region in (5.7) can have at most one corresponding constraint of type a) in (5.4).*

- *Each constraint of type a) has at least one corresponding full-dimensional critical region.*

- *If critical region s has a corresponding constraint j, it holds that $F_s = E_j^{\{u,l\}}$ and $f_s = W_j^{\{u,l\}}$.*

Table 5.1: Complexity of different solution strategies for QP (5.4). The basic and the efficient explicit algorithm together with a corresponding complexity evaluation can be found in (Borrelli et al., 2015).

algorithm	number of flops (worst case)	storage demand (real numbers)
basic explicit	$2nN_c \geq 2n(n+1)(N_l + N_u)$	$(n+1)N_c \geq (n+1)^2(N_l + N_u)$
efficient explicit	$(2n-1)n_{CR} + N_c \geq 3n(N_l + N_u)$	$(n+1)n_{CR} \geq (n+1)(N_l + N_u)$
Algorithm 13	$(2n+1)(N_l + N_u)$	$(n+1)(N_l + N_u)$
Algorithm 14	$(2n+1)(\max\{N_l, N_u\} + 1)$	$(n+1)(N_l + N_u) + n$

- *It holds that $N_l + N_u \leq n_{CR}$.*

Proof. As the critical regions are full-dimensional, they define the affine mapping $z^*(x)$ for x in the critical region uniquely. With the fact that both representations are equivalent, for each critical region either the given equalities have to hold or $F_s = 0$, $f_s = 0$. If one critical region had more than one corresponding type a) constraint, all these constraints would have to be equal as the affine mapping associated to a critical region is defined uniquely. All except for one of these identical constraints would be redundant and could be removed. Each type a) constraint is active for a full-dimensional set of states and, thus, equivalence of the representations requires existence of a corresponding full-dimensional critical region. The first three claims imply that there exists a surjective mapping from critical regions to type a) constraints defined for all critical regions with $F_s \neq 0$, $f_s \neq 0$. This implies the last claim. □

Evaluating the explicit solution (5.7) for a given state x, generally the numerically most expensive part is to identify the critical region $\mathcal{CR}_s$ which contains x. Various efficient strategies to solve this point location problem have been proposed. We compare the numerical complexity of a basic and an efficient point location strategy presented in (Borrelli et al., 2015) to the complexity of applying Algorithm 13 or 14. To enable a meaningful comparison, we use the last claim of Lemma 5.8 and the following relation. A full-dimensional polytope in n-dimensional space comprises at least $n+1$ constraints, i.e., $c_s \geq n+1$. Thus, the total number of constraints N_c present in the explicit solution can be lower bounded as $N_c \geq (n+1)n_{CR} \geq (n+1)(N_l + N_u)$. The computational complexities for the algorithms presented in (Borrelli et al., 2015) are given in Table 5.1 together with (generally conservative) lower bounds in terms of N_l and N_u.

Algorithm 13 requires less than $(2n+1)/3n$ times the worst case computations of the efficient explicit scheme, i.e., for $n > 1$ Algorithm 13 is guaranteed to be simpler than the efficient explicit scheme comparing worst cases. The guaranteed advantage grows with growing state space dimension n and can, due to the conservative bound on N_c used, be generally expected to be even larger. Beyond that, for the typical cases of $N_l = N_u$ and $n \ll N_l$, Algorithm 14 saves almost half of the computations of Algorithm 13.

Finally, the last column of Table 5.1 contains information on storage requirements of the different algorithms. For all algorithms storage requirement is very similar except for the basic explicit solution strategy, which requires considerably more data to be stored. Note that, whereas Algorithms 13 and 14 directly yield the optimizer, the basic and the efficient explicit algorithm compared here only identify the critical region containing the given state.

Table 5.2: Average percentage of different constraint types observed in the example.

type a)	type b)	type c)
81 %	18 %	1 %

In order to obtain the corresponding optimizer, additional $2n + 1$ flops are required and additional $(n + 1)n_{\mathcal{CR}}$ numbers need to be stored.

5.4.2 Examples

Numerical example

In order to evaluate the complexity of the almost explicit solution approach for an actual example, let us reconsider the oscillating masses example introduced in the previous chapter.

Example 5.9 (Example 4.29 continued). *In Section 4.5.1 parametrizations using the values $q = 1$ and $K \in \{2, 5, 15\}$ were computed for this example. As in all cases $q = 1$ was chosen and the parametrizations were determined via the third approach to find feasible parametrizations, both the almost explicit and the simplified almost explicit solution strategy are applicable to solve the resulting parametrized optimization problems. In the following, we evaluate the results in this respect.*

Summing over the parametrizations for all values of K, a total of 22 parametrized mpQPs results. For each of them, we first prepared all data required for the almost explicit and the simplified almost explicit solution strategy. This comprises classification of the constraints and preparation of the input data for Algorithm 13 and 14, respectively. In all cases this required at most few seconds rendering these preparations orders of magnitude faster than computation of the corresponding explicit solutions via generic methods.

Typically only single type c) constraints were present whereas most of the constraints were of type a). Neglecting type d) constraints, the fractions of the constraint types observed averaged over all parametrized optimization problems are given in Table 5.2.

Regarding the worst case online complexity, the bounds in Table 5.1 guarantee for this problem (due to its eight-dimensional state space) that the almost explicit solution approach requires less than $(2n+1)/(3n) \approx 70.8\%$ of the number of flops the efficient explicit scheme requires. Based on the computed data and on explicit solutions computed for the parametrized problems, Table 5.1 can also be evaluated non-conservatively. It turned out that in all 22 cases the advantage of the almost explicit strategy over the efficient explicit strategy was much larger than estimated via the bounds in Table 5.1. On average the worst case complexity for the efficient explicit strategy was more than 5 351 flops whereas for the almost explicit strategy this was less than 1 744 flops. The fraction of worst case computations needed in the almost explicit scheme is on average 27.5 % of the worst case computations required in the efficient explicit approach. For the simplified almost explicit strategy, in fact worst case computations could be almost halved with respect to the almost explicit strategy. All numbers are given in Table 5.3.

Note that for this problem any move-blocking strategy would require at least a $q = 2$-dimensional decision variable as the system input is $m = 2$-dimensional. So the almost explicit solution strategy is not applicable in a move-blocking based MPC scheme in this

Table 5.3: Complexity of different solution strategies for oscillating masses example, averaged over all parametrized mpQPs. The efficient explicit algorithm is taken from (Borrelli et al., 2015).

algorithm	number of flops (worst case, averaged)	normalized complexity (worst case, averaged)
efficient explicit	5 351.5	100 %
Algorithm 13	1 743.3	27.5 %
Algorithm 14	888.6	14.0 %

case and application of the semi-explicit approach is necessary to enjoy the benefits of the almost explicit solution strategy.

Application example

Due to the simplicity of all involved online operations, almost explicit MPC is attractive for practical applications and the low requirements regarding computational power and data to be made available during run-time make it particularly suitable for the implementation on low cost hardware and embedded systems. This was validated experimentally in a Master's thesis project (Schmidt, 2016), where an almost explicit MPC scheme was implemented on low-cost embedded hardware to balance a two-wheeled mobile robot. The robot was built from LEGO Mindstorm NXT according to the one introduced in (Yamamoto, 2008) and also treated in an MPC context in (Zometa et al., 2012), see Figure 5.2 for the robot. Here, only the pitch dynamic was considered so that the system could be modeled using four states and a scalar input. State, input and terminal constraints were taken into account. Feasible parametrizations with a scalar parameter ($q = 1$) were computed via the second approach (Algorithm 9). The obtained controller was then implemented and run on the actual system.

All online algorithms could be programed in a straight forward manner thanks to the simple elementary operations they consist of. The amount of data to be stored turned out to be well below the storage capabilities of the low-cost hardware used and it scaled favorably in the prediction horizon length N. See Figure 5.3 for an illustration of the observed storage requirements. Likewise, despite the restricted computational power of the employed microcontroller, the almost explicit MPC scheme could be executed at rather high sampling rates in the order of milliseconds and, in fact, stabilized the robot. For more details on the experimental setup, the implementation details and the obtained results, we refer to (Schmidt, 2016).

In conclusion, these experimental results verified that the almost-explicit MPC scheme is well applicable to control an actual mechatronic system using very limited computational resources. On the other hand, the results are still to be seen only as a proof of concept since a scheme close to the nominal one has been considered[2] and typical simplifications were left unexploited (e.g. omitting terminal constraints, sacrificing guarantees, application of

---

[2] The main difference with respect to the nominal semi-explicit MPC schemes was to omit the cost comparison which is used to enforce a decrease in the predicted open-loop costs. As is to be expected using an inaccurate model, in the experiments this resulted in too large preference of the candidate input sequence and thereby deteriorated control performance.

Figure 5.2: Robot built from LEGO Mindstorms NXT and balanced running almost explicit MPC algorithm on low-cost embedded hardware. Figure taken from (Schmidt, 2016).

the simplified almost explicit scheme, exploitation of symmetry, etc.). These simplifications and further tailoring the implementation would certainly extend applicability to models with more states and inputs and to achieving even higher sampling rates.

5.4.3 Evaluation and discussion

Let us next briefly evaluate the results of this chapter theoretically. The optimization problem considered is a very special case of a parameter-dependent optimization problem: A convex multi-parametric quadratic program in a scalar decision variable. Numerical solution of this problem for a fixed value of the parameter is generally very simple. This is, depending on the perspective, either because the search direction for the optimizer is obtained trivially or because the problem has generally few active sets which each consist of at most one element, i.e., one active constraint determines the optimizer uniquely. In order to exploit these circumstances best, a solution strategy was tailored to this problem class. This solution strategy requires to prepare a priori some quantities offline and in turn consists of a very simple online procedure with totally predictable complexity. The a priori preparations proposed here are generally much simpler than computation of the explicit solution and also the solution procedure itself can be guaranteed to be simpler in the worst case than evaluation of the explicit solution for state dimensions greater than one. As the parametric optimization is in fact not solved completely offline, but only some data is prepared to simplify online solution of the optimization, the approach again fulfills the criteria of the suggested "semi-explicit methods" (Alessio and Bemporad, 2009): "... *semi-explicit methods should be also sought, in order to pre-process of line as much as possible of the MPC optimization problem without characterizing all possible optimization outcomes, but rather leaving some optimization operations on-line.*"

The proposed approach has slight similarities with the simplified explicit MPC approaches suggested in (Borrelli et al., 2010; Kvasnica et al., 2015). Therein, the idea is to not store the critical regions of the explicit solution explicitly but to make available analytical expressions

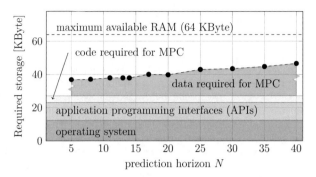

Figure 5.3: Illustration of the contributions to overall program size to run almost explicit MPC on the LEGO Mindstorm NXT robot. The code for the almost explicit MPC scheme possesses marginal size, the required data for the almost explicit MPC scheme scales well in the prediction horizon. Overall, the total size of the program code is well below the maximum available RAM. Modified figure taken from (Schmidt, 2016).

of the (primal and) dual variables for all optimal active sets. Identifying the corresponding active set for a given state is then simplified due to the simplified evaluation of primal and dual feasibility. In the almost explicit approach, the critical regions are not stored explicitly, either. Once the relevant constraints for primal feasibility have been identified, only three active set candidates remain, of which the optimal one can be identified comparing the corresponding objective function values. Beyond that, interpreting the identification of the relevant constraints as solving two linear programs with convex optimal value function, connections to the efficient point location algorithm for multi-parametric linear programs found in (Borrelli et al., 2015) can be seen.

As the almost explicit solution strategy addresses a rather specific problem class, its applicability is generally limited. It is directly applicable to (non-parametrized) MPC schemes only in the unlikely case of having a scalar system input and using prediction horizon $N = 1$. Applicability can in principle be extended applying parametrizations in an MPC scheme. Yet, only very few parametrizations enable to reduce the decision variable to a scalar and still achieve a satisfactory level of control performance and feasible set. Besides some interpolation based schemes (see e.g. (Rossiter et al., 2004)), the semi-explicit MPC approach is a notable exception from that. This circumstance closely links the almost explicit solution strategy and the semi-explicit MPC approach. Conversely, the significant numerical and structural simplicity of the almost explicit optimization even more justifies the additional effort required to handle a piecewise defined parametrization. In this sense, the simplicity of the almost explicit solution strategy is a strong argument for semi-explicit MPC schemes. As mentioned above and shown in the application example, almost explicit MPC is readily applicable to control problems which require fast sampling rates employing low-cost computational hardware.

5.5 Summary and extensions

5.5.1 Summary

In this chapter we addressed the numerical solution of the parametrized optimization in the special case of a linear MPC problem with quadratic objective functions and using univariate parametrizations. We formulated an extremely simple case analysis which solves the parametrized optimization for a given state in this situation. A first additional simplification of this scheme was achieved by classifying different types of constraints in the optimization and removing non-required ones in advance. A second simplification beyond that was possible if a parametrization with independent feasibility part is used as this typically allows to exclude half of the cases from the case analysis in advance. A detailed theoretical complexity analysis of the proposed algorithms was executed. It proved that the proposed algorithms have reduced worst case complexity compared to the online evaluation of the corresponding explicit solution for systems with more than two states. A numerical example from the previous chapter was revisited and an evaluation in view of the almost explicit solution strategy revealed that advantages in fact are considerably larger. All results of this chapter are directly applicable in the semi-explicit MPC framework, reducing the amount and complexity of online operations required therein further such that the approach is made even more attractive.

5.5.2 Extensions

Some extensions of the almost explicit approach are immediate. Proposition 5.1 indicates how to generalize the approach to objective functions beyond quadratic ones. Most relevant, for linear objective functions an optimizer is always at the upper or lower bound depending on the sign of the corresponding coefficient in the objective function. Consequently, the numerical procedure can be simplified to evaluating either upper or lower bounds, similar to the simplified almost explicit approach.

As illustrated via the complexity comparison, the proposed approach has very low worst case complexity. On the downside, the actual complexity will generally not be below the worst case complexity. Efforts to improve the balance of these aspects could be made. Similar to as it was done in the simplified almost explicit scheme, additional hyperplanes could be used so that for states at either side of the hyperplane only a subset of the upper and lower bounds can become active. Correspondingly, only a subset of the constraints would have to be evaluated.

Taking into account that the almost explicit solution strategy is to be applied in a semi-explicit MPC scheme, further adaptations are possible. In the general semi-explicit approach for linear systems we discussed how, due to overlapping feasible regions of the single parts of a parametrization, explicit solutions of the corresponding parametrized mpQPs could be simplified by removing some of the critical regions from the complete solution. In the almost explicit case, critical regions correspond to type a) upper or lower bounds. So, if a specific scalar mpQP is known a priori to be evaluated only for a certain subset of all feasible states, it might be that some of the constraints cannot become active and could be removed beforehand. For example, if the mpQP corresponding to Figure 5.1 is evaluated only for positive states x, the left type a) constraint can never become active and could be omitted at no losses.

Beyond these aspects, the scalar decision variable which is used in the almost explicit MPC scheme can simplify adaptation of the scheme to using elliptic terminal constraints. An elliptic terminal constraint translates into an upper bound on the squared scalar decision variable. Computing and using a piecewise affine upper bound on this scalar function will generally be simpler than computing and using an inner approximation of the elliptic terminal set in a high-dimensional state space. Numerical tests are necessary to evaluate if a sufficiently good piecewise upper bound defeats the goal of overall simplification when replacing polytopic by elliptic terminal constraints.

The current chapter concludes our focus on applying semi-explicit MPC to linear systems and exploiting the achievable simplifications. We are now ready to investigate the more general case of applying semi-explicit MPC to certain nonlinear systems more in detail in the next chapter.

Chapter 6

Semi-explicit MPC for nonlinear systems

In Chapter 3 of this thesis we introduced and evaluated the semi-explicit MPC approach for general nonlinear systems. Yet, therein we did not address the issue of finding a parametrization such that the parametrized optimization is guaranteed to be feasible. As the capability of handling constrained control problems is a key feature and selling point of model predictive control, it is a central requirement to guarantee feasibility of the nonlinear semi-explicit MPC scheme even when general constraints are present. The current chapter primarily tackles this requirement: Given a rather general nonlinear system with corresponding MPC problem, how can a suitable parametrization be found such that the parametrized optimization is guaranteed to be feasible for a given set of states?

This chapter is partly based on (Goebel and Allgöwer, 2014b).

6.1 Introduction, problem formulation and preliminaries

6.1.1 Introduction and problem formulation

The general problem setup is again inherited from previous chapters just as all basic assumptions which are also kept throughout this chapter. A nonlinear system subject to state and input constraints is considered. This system is to be controlled such that all constraints are satisfied, a given stage cost is minimized along closed-loop trajectories and the origin is asymptotically stabilized. To this end, a setpoint stabilizing MPC scheme is to be applied which comprises a terminal cost and terminal constraints which together with a known terminal control law fulfill the stability implying Assumption 3.1. As discussed above, this results in the finite horizon open-loop optimal control problem formulated in Optimization (2.2).

Our goal in the current chapter is to formulate a semi-explicit MPC scheme which simplifies the solution of the latter problem and maintains system theoretic guarantees. In particular, we address offline computation of the ingredients of the semi-explicit MPC scheme which consists of a parametrization which is guaranteed to be feasible for a given set of states.

We start by considering a general nonlinear discrete-time system $x^+ = f(x, u)$. For some of the results below, we will then introduce and make use of further assumptions on f. As the considered system class is very general, a generic solution approach to such problem will generally not be optimal for a given specific problem but rather introduce some conservatism. It is, nevertheless, the goal of this chapter to present generally applicable methods. The results are not as complete as those presented for linear systems and are thus also to be seen as an outlook and starting point for future research or more specific considerations for a given control problem.

6.1.2 Preliminaries

Whereas following the MPC paradigm it is the goal to find and apply inputs which are admissible and optimal with respect to Optimization (2.2), we will in the current chapter propose an approach which online does not directly compute solutions thereof. Instead, an auxiliary transformed optimization problem will be used as a tool to find input sequences which are easily transformable into solutions of the original problem. In the following, we will introduce the transformed problem as well as its parametrized version and elaborate on the connection of solutions to the original and the transformed problem.

Application of a transformed problem becomes necessary as (at least parts of) the results in the current chapter are based on formulating the MPC problem with transformed input

$$v = u - K_p x \qquad (6.1)$$

as decision variable. Therein x is the current system state and $K_p \in \mathbb{R}^{m \times n}$ is a chosen matrix. To simplify the notation, we will cast the input transformation via a transformed system dynamic function $\tilde{f}$, which fulfills

$$x^+ = \tilde{f}(x, v) = f(x, K_p x + v). \qquad (6.2)$$

Such input transformation is typically used to pre-stabilize a system, i.e. such that the origin of $x^+ = \tilde{f}(x, 0)$ is an asymptotically stable equilibrium of the system even if the origin of $x^+ = f(x, 0)$ is not stable. Here it will be used more generally to achieve that system $\tilde{f}$ has improved properties over system f for the considered purposes as will be detailed below. The finite horizon open-loop optimal control problem corresponding to Optimization (2.2) but formulated in terms of v inputs is then given by

$$
\begin{aligned}
\min_{\mathcal{V}} \quad & \sum_{j=0}^{N-1} \ell(x_j, K_p x_j + v_j) + V_T(x_N) \\
\text{s.t.} \quad & x_{j+1} = \tilde{f}(x_j, v_j), && j = 0, \ldots, N-1 \\
& x_0 = x, \\
& x_j \in \mathbb{X},, && j = 1, \ldots, N-1 \\
& K_p x_j + v_j \in \mathbb{U}, && j = 0, \ldots, N-1 \\
& x_N \in \mathbb{X}_T
\end{aligned}
\qquad (6.3)
$$

where $\mathcal{V} = [v_0^\top, \ldots, v_{N-1}^\top]^\top$. This problem is equivalent to the original Optimization (2.2) formulated for the non-transformed system f in the sense that if the sequence $v_0, \ldots, v_{N-1}$ is admissible and optimal with respect to (6.3), the input sequence $K_p x_0 + v_0, \ldots, K_p x_{N-1} + v_{N-1}$ is admissible and optimal with respect to (2.2) and both problems yield the same objective function value. Thus, solutions of (6.3) can easily be transformed into solutions of (2.2) and vice versa.

In order to establish system theoretic guarantees for the semi-explicit MPC scheme, it was crucial to be able to compute from a given solution of Optimization (2.2) for initial state x a solution for the successor state x^+ which yields a sufficiently low objective function value. For Optimization (2.2) such candidate solution is given by $U_C = [u_1^\top, \ldots, u_{N-1}^\top, \kappa(x)^\top]^\top$. Transforming this into v inputs yields $\mathcal{V}_C = [u_1^\top - (K_p x_1)^\top, \ldots, u_{N-1}^\top - (K_p x_{N-1})^\top, \kappa(x)^\top - (K_p x_N)^\top]^\top = [v_1^\top, \ldots, v_{N-1}^\top, \kappa(x)^\top - (K_p x_N)^\top]^\top$. This is as corresponding candidate solution to the transformed problem (6.3). Thus, the standard procedure to generate a candidate input sequence can also be applied for the transformed problem if a transformed terminal control law $\tilde{\kappa}(x) = \kappa(x) - K_p x_N$ is used. Being equivalent to the non-transformed candidate solution, the transformed candidate solution yields the same cost and consequently cost decrease. Thus, in summary, Optimization (6.3) provides basically the same features as Optimization (2.2) which were exploited to formulate a semi-explicit MPC scheme with guaranteed stability.

Hence, it is the strategy here to parametrize the decision variable $\mathcal{V}$ of Optimization (6.3) in order to simplify the numerical solution of this optimization and, overall, formulate a semi-explicit MPC scheme based on this optimization.

Before we formulate the latter optimization in its parametrized form, we introduce a generalization of the parametrization which is made in this chapter. We relax the requirement, that the parametrization has to be evaluated for the current system state (i.e. $p(x, \cdot)$ has to be used if the current state is x) but generalize to using $p(\tilde{x}, \cdot)$ for $\tilde{x}$ a fixed state close to x and chosen depending[1] on x. This will be explained more in detail below such that also the virtue of this measure becomes obvious.

In order to formulate the parametrized transformed optimization problem, we define $p^j(\cdot, \cdot)$ for $j = 0, \ldots, N-1$ to be the j-th predicted input v_j via part p_i of the parametrization, to be more precise $p^j(x, \tilde{U})$ are the elements $1 + jm, \ldots, (j+1)m$ of the vector $p_i(x, \tilde{U})$. Most of the following considerations have to be done for each part $p_i, i \in \{1, \ldots, K\}$ of the parametrization independently. Hence, we omit the index i where it is not needed in order to simplify notation. Using the latter definition, the parametrized version of problem (6.3) is given by

$$
\begin{aligned}
\min_{\tilde{U}} \quad & \sum_{j=0}^{N-1} \ell(x_j, K_p x_j + p^j(\tilde{x}, \tilde{U})) + V_T(x_N) \\
\text{s.t.} \quad & x_{j+1} = \tilde{f}(x_j, p^j(\tilde{x}, \tilde{U})), && j = 0, \ldots, N-1 \\
& x_0 = x, \\
& x_j \in \mathbb{X}, && j = 1, \ldots, N-1 \\
& K_p x_j + p^j(\tilde{x}, \tilde{U}) \in \mathbb{U}, && j = 0, \ldots, N-1 \\
& x_N \in \mathbb{X}_T.
\end{aligned}
\tag{6.4}
$$

where the same quantities as introduced in (2.2) are used. It will be the goal in this chapter to find parametrizations such that the latter optimization is feasible, more in detail to find for a given desired feasible set $\mathcal{X}_f \subset \mathbb{R}^n$ sets $\{\hat{\mathcal{D}}_1, \ldots, \hat{\mathcal{D}}_K\}$ such that $\mathcal{X}_f \subseteq \hat{\mathcal{D}}_1 \cup \cdots \cup \hat{\mathcal{D}}_K$ and (6.4) is feasible for $x \in \hat{\mathcal{D}}_i$.

Before we address this problem, we will look more in detail at the correspondence of the

---

[1] Note that if continuity of the parametrization at the origin is required (see Proposition 3.8), the mapping $x \mapsto \tilde{x}$ has to be chosen suitably.

transformed and the untransformed optimization problem. We introduce the notation

$$x_{j+1}(x, \tilde{x}, \tilde{U}) = f(x_j(x, \tilde{x}, \tilde{U}), u_j(x, \tilde{x}, \tilde{U})), \tag{6.5}$$

$$u_j(x, \tilde{x}, \tilde{U}) = K_p x_j(x, \tilde{x}, \tilde{U}) + p^j(\tilde{x}, \tilde{U}), \quad j = 0, \dots, N-1, \tag{6.6}$$

for the N step predicted state sequence from initial condition x and the corresponding input sequence to f resulting from the transformed input sequence $v_0 = p^0(\tilde{x}, \tilde{U}), \dots, v_{N-1} = p^{N-1}(\tilde{x}, \tilde{U})$, respectively. Therein, $x_0(x, \tilde{x}, \tilde{U}) = x$ holds. It is clear that by successively eliminating the states in (6.6) via (6.5), the predicted input sequence in u coordinates can be expressed in terms of x and $\tilde{U}$. The result can be interpreted as a transformed parametrization suitable to approximate inputs in the original u coordinates. In the linear case this results in a parametrization of the same structure as the original one, see the end of Subsection 4.2. In the nonlinear case, this is not as straight forward as the linear input transformation (6.1) is propagated trough nonlinear dynamics. Thus, we stick to working in the regime of transformed inputs keeping in mind that in principle a transformation back into the original inputs is possible and that required properties of the original optimization carry over to the transformed one.

For the remainder of this chapter, we will focus on the offline part of the semi-explicit MPC scheme for nonlinear systems. First, we will formulate feasibility tests for the parametrized optimization problem under different assumptions on the system dynamic function and the parametrization used. These results will then be turned into an offline algorithm to compute feasible parametrizations. Finally, we will evaluate the method in an illustrative numerical example as well as theoretically.

6.2 Guaranteeing feasibility for nonlinear systems

The fundamental challenge when assessing or trying to achieve feasibility of the parametrized optimization for a general non-convex problem is to conclude based on a finite number of tests on feasibility for an infinite number of different states. In the case of linear systems, convexity of the problem made it sufficient to verify feasibility at vertices of a convex state set in order to conclude feasibility for all states within the convex set. Here, we specifically do not assume convexity which means that different techniques are required. Next, we will first formulate a feasibility test based on robust optimization and robust MPC techniques and then below employ these results in an iterative procedure which can render parametrizations feasible.

The general approach to test feasibility for states x in a given set $\hat{\mathcal{D}} \subset \mathbb{R}^n$ is as follows. Choose a set $\Delta \subset \mathbb{R}^n$ and test states $\{z_1, \dots, z_{n_t} \in \mathbb{R}^n\}$ such that

$$\hat{\mathcal{D}} \subseteq \Delta \oplus \{z_1, \dots, z_{n_t}\} \tag{6.7}$$

holds. Then, for each test state z_r a test problem is formulated which if feasible ensures that the parametrized optimization problem is feasible for $x \in \{z_r\} \oplus \Delta$. If all test problems are feasible, overall feasibility for $x \in \hat{\mathcal{D}}$ follows.

6.2.1 Feasibility test as robust optimization problem

The requirements formulated regarding the "test problem", in fact, exactly describe a robust optimization problem. In robust optimization, it is the goal to solve a parameter-dependent optimization problem such that the solution is valid not only for a nominal parameter value but for a whole set of parameter values. More formally, the following problem is a robust optimization problem and summarizes the requirements for the test problem stated above. Solve for given $z \in \mathbb{R}^n$ a problem of the form

$$\min_{\tilde{U}} J(z, p(z, \tilde{U}))$$
$$\text{s.t. } g(x, p(x, \tilde{U})) \leq 0 \text{ for all } x \in \{z\} \oplus \Delta. \tag{6.8}$$

Solving such problem in a general case is at least challenging and a typical strategy in some simpler cases is to work based on a so-called robust regularization of the constraints. A robust regularization is a function tightly upper bounding the constraint function for all considered values of the parameter. In our case, this would require to find mappings $\overline{g}_i(z, p(z, \tilde{U})) = \sup_{x \in \{z\} \oplus \Delta} g_i(x, p(x, \tilde{U}))$ for each element i of the vector valued constraint function g. Replacing the constraints in (6.8) by their robust relaxations clearly yields an equivalent non-robust optimization problem. Due to the generality of the considered problem, we cannot assume that a robust regularization can be found and the best one could hope for is to find a conservative upper bound on the robust regularization. This could be done based on rather general properties of the constraint function and the set Δ, for example via interval arithmetic or based on determining and using Lipschitz constants. Instead of following this route further, we have a closer look at upper bounding the constraint function based on system theoretic insights in the next subsection.

6.2.2 Feasibility test via system theoretic properties

Testing feasibility based on system theoretic properties is in principle similar to the robust optimization based approach, but, in addition system theoretic insights are exploited. Loosely speaking, given a set of initial conditions and an input sequence or a set of input sequences, the set of reachable states propagating the initial conditions through the system dynamic via the inputs is computed or bounded. These sets of reachable states is then used to conclude about feasibility of the parametrized optimization problem directly or is used to pre-tighten the constraints in a feasibility test. This pre-tightening is somewhat comparable to the robust regularizations used in the robust optimization approach.

The idea to bound the sets of reachable states based on system theoretic properties clearly is inspired by results which have been developed and applied for similar related problems as, for example, robust MPC and nonlinear explicit MPC. Available results in these areas differ among others in how bounds on the reachable sets are computed and in whether or not the reachable sets are computed directly or in form of so-called *tubes* around nominal sequences. *Interval arithmetic* was used to compute reachable sets in (Limon et al., 2005) in a robust nonlinear MPC setting and in (Summers et al., 2010) in the context of explicit MPC for nonlinear systems. *Zonotopes* were used in (Bravo et al., 2006) to bound the sets of reachable states in robust nonlinear MPC. A technique called *DC programming* is used for state estimation to tightly bound sets of possible states in (Alamo et al., 2008) and its application for nonlinear explicit MPC was suggested in (Raimondo et al., 2012). *Lipschitz continuity* of the system dynamic function was used in (Limon Marruedo et al., 2002) to bound the effect of additive disturbances in MPC and it was used in (Aguilera and Quevedo, 2011) to cope with discretization of the input in MPC. *Incremental stability* of the system was used in (Bayer et al., 2016) to establish an explicit MPC scheme for nonlinear systems.

Results along the lines of all of these approaches could in principle be developed for the considered purposes. Next, we first consider a general level set-based approach for the problem at hand and then show that results along the lines of the latter two approaches (based on Lipschitz continuity and incremental stability) are covered as special cases.

Let us first elaborate a little more on the general avenue we follow here. Again, we work based on test states $\{z_1, \ldots, z_{n_t}\}$ and sets $\{z\} \oplus \Delta$ around the test states. The idea is then to find for each test state z a parameter $\tilde{U}^*$ such that the parametrized input sequence $p(z, \tilde{U}^*)$ does not only drive the test state z to the terminal set subject to all constraints but which also ensures that $p(\tilde{x}, \tilde{U}^*)$ drives any $x \in \{z\} \oplus \Delta$ into the terminal set subject to all constraints for a suitable choice of $\tilde{x} \in \{x, z\}$. The choice of $\tilde{x} \in \{x, z\}$ will introduce flexibility in the design and will be made below. To this end, $\tilde{U}^*$ has to be found based on state, terminal and input constraints which have been tightened with respect to the original ones. This tightening is computed based on an a priori bound on the distance of the state sequences originating at z and x, respectively, and of the corresponding input sequences $p(z, \tilde{U}^*)$ and $p(\tilde{x}, \tilde{U}^*)$, respectively. Summarizing, the test states z have to be steered into the terminal set such that also a tube around the test state sequence which contains all sequences initialized in $\{z\} \oplus \Delta$ satisfies all constraints. This idea is formalized in the following result.

Lemma 6.1 (Feasibility test). *Let $\tilde{x} \in \{x, z\}$ be fixed and let for a given set $\Delta \subset \mathbb{R}^n$ there be sets $\Omega_1^x, \ldots, \Omega_N^x \subset \mathbb{R}^n$ and $\Omega_0^u, \ldots, \Omega_{N-1}^u \subset \mathbb{R}^n$ such that for all $x \in \{z\} \oplus \Delta$ and for all $\tilde{U} \in \mathbb{R}^q$*

$$x_j(z, z, \tilde{U}) - x_j(x, \tilde{x}, \tilde{U}) \in \Omega_j^x, \ j = 1, \ldots, N, \tag{6.9}$$

$$u_j(z, z, \tilde{U}) - u_j(x, \tilde{x}, \tilde{U}) \in \Omega_j^u, \ j = 0, \ldots, N-1, \tag{6.10}$$

holds. If the problem

$$\begin{aligned}
\min_{\tilde{U}} \ & 0 \\
\text{s.t.} \ & x_{j+1} = \tilde{f}(x_j, p^j(z, \tilde{U})), && j = 0, \ldots, N-1 \\
& x_0 = z, \\
& x_j \in \mathbb{X} \ominus \Omega_j^x, && j = 0, \ldots, N-1 \\
& K_p x_j + p^j(z, \tilde{U}) \in \mathbb{U} \ominus \Omega_j^u, && j = 0, \ldots, N-1 \\
& x_N \in \mathbb{X}_T \ominus \Omega_N^x.
\end{aligned} \tag{6.11}$$

is feasible then problem (6.4) *is feasible for all $x \in \{z\} \oplus \Delta \cap \mathbb{X}$.*

The sets Ω_j^x and Ω_j^u therein can be interpreted as cross sections of the tubes around the state and input sequences corresponding to the test state z which contain the sequences corresponding to an actual state x.

Finding tube cross sections

In order to apply the latter lemma, it remains to find the tube cross sections Ω_j^x and Ω_j^u which bound the distance of the state sequences and of the input sequences. The strategy we look at more in detail here is to use level sets of suitable functions to overestimate these distances and to use information on the dynamic evolution of such level sets. The following assumption specifies the information we suppose to be available.

Assumption 6.1. *Let functions $V_1 : \mathbb{R}^n \to \mathbb{R}$ and $V_2 : \mathbb{R}^m \to \mathbb{R}$ be known such that suitable class $\mathcal{K}_\infty$ functions $\underline{\alpha}$, $\overline{\alpha}$ and α satisfy for all $e_x \in \mathbb{R}^n$ and all $e_u \in \mathbb{R}^m$*

$$\underline{\alpha}(|e_x|) \leq V_1(e_x) \leq \overline{\alpha}(|e_x|) \quad \text{and} \quad V_2(e_u) \leq \alpha(|e_u|) \tag{6.12}$$

and the following assumptions are satisfied.

a) For all $z, x \in \mathbb{X}$ and all v_z, v_x with $K_p z + v_z \in \mathbb{U}$, $K_p x + v_x \in \mathbb{U}$ the relation

$$V_1 \left(\tilde{f}(z, v_z) - \tilde{f}(x, v_x) \right) \leq L_1^x V_1(z - x) + L_1^v V_2(v_z - v_x) \tag{6.13}$$

is valid with known constants L_1^x and L_1^v.

b) For all $z, x, \tilde{x} \in \mathbb{X}$ and $\tilde{U} \in \mathbb{R}^q$ the relation

$$V_2 \left(u_j(z, z, \tilde{U}) - u_j(x, \tilde{x}, \tilde{U}) \right) = V_2 \left(N^j(\phi(z) - \phi(\tilde{x})) + K_p(x_j(z, z, \tilde{U}) - x_j(x, \tilde{x}, \tilde{U})) \right)$$
$$\leq L_2^{x_0} V_1(z - x) + L_2^x V_1 \left(x_j(z, z, \tilde{U}) - x_j(x, \tilde{x}, \tilde{U}) \right)$$
$$\tag{6.14}$$

is valid for $j = 0, \ldots, N - 1$ with known constants $L_2^{x_0}$ and L_2^x.

Recall that ϕ denotes the state-dependent part of the parameterization. By N^j we furthermore denote the j-th part of the predicted sequence. The first part of the assumption solely concerns the system dynamics and means that given two states which differ by e_x and propagating the states through the system $\tilde{f}$ with two inputs which differ by e_u, the successor states will differ by e^+ where e^+ is contained in the level set $\Omega_{j+1}^x = \{x | V_1(x) \leq L_1^x V_1(e_x) + L_1^v V_2(e_u)\}$. The second part of the assumption concerns the parametrization and the input transformation term $K_p x$. It means that given two states which differ by e_x the difference e_u^j of the corresponding predicted input sequences at time-step j is contained in the level set $\Omega_j^u = \{u | V_2(u) \leq L_2^{x_0} V_1(e_x) + L_2^x V_1(e_x^j)\}$ where e_x^j is the difference of the predicted state sequences at time step j.

In principle, (6.13) and (6.14) can now be used to find (level) sets Ω_j^x and Ω_j^u which contain the elements of the difference of the predicted state e_x^j and input sequences e_u^j at time step j and which depend on the difference of the initial states x and z. This is achieved via iteratively executing the following procedure for increasing value of j covering all elements of the predicted sequences. First, the expression for the set containing e_u^j is plugged into the relation for e_x^j to find a level set for e_x^j depending on the level set containing e_x. This result is then used to find an expression for e_u^j depending on the level set containing e_x. Eventually, all level sets are expressed depending on the constants L_1^x, L_1^v, $L_2^{x_0}$ and L_2^x and of the initial level set $V_1(e_x)$. The resulting sets can then be applied to formulate a feasibility test (6.11) according to Lemma 6.1.

The generality in which the results have been presented so far will render these calculations tedious with little additional insights and limited applicability. Thus, we next consider two simplified meaningful manifestations thereof in more detail. For these special cases explicit expressions of the level sets will be derived.

Tube cross sections for Lipschitz continuous systems

An immediate interpretation of Assumption 6.1 is obtained if V_1 and V_2 therein additionally fulfill the requirements of a norm. In this case, the constants L_1^x, L_1^v, $L_2^{x_0}$ and L_2^x can be taken as the Lipschitz constants of the respective mapping with respect to the metrics induced by the norms. More in detail, L_1^x and L_1^v can be taken as the Lipschitz constants of the system dynamic function $\tilde{f}$ of its first and second argument, respectively. The constant $L_2^{x_0}$ can be taken as the Lipschitz constant of the parametrization p_i with respect to the state x and L_2^x can be the Lipschitz constant of the linear mapping $x \mapsto K_p x$. The assumption then follows from upper bounding the left hand sides of (6.13) and (6.14) first based on the triangular inequality (which holds as V_1 and V_2 are norms) and then based on the Lipschitz continuity relations. Summarizing, if the system dynamic function $\tilde{f}$ is Lipschitz continuous on its domain with known Lipschitz constants, part a) of Assumption 6.1 is fulfilled. The constant $L_2^{x_0}$ exists for a suitable choice of ϕ in the parametrization and L_2^x exists as linear mappings are Lipschitz continuous. Existence of these constants implies part b) of the assumption.

Another implication of assuming V_1 and V_2 to be norms is that they fulfill the homogeneity property which drastically simplifies scaling of the involved level sets. Scaling of a level set by a factor in this case simply means to scale each element of the level set by the same factor, i.e., $\{x|V_1(x) \leq k_1 k_2\} = \{k_1 x | V_1(x) \leq k_2\}$ for $k_1, k_2 > 0$.

In order to formulate a compact result for Lipschitz continuous systems next, we additionally assume here that no input transformation is applied, i.e. $K_p = 0$ which implies $\tilde{f} = f$ and $L_2^x = 0$.

Proposition 6.2 (Tube cross sections for Lipschitz continuous system)*. Let V_1 and V_2 be norms defined on $\mathbb{R}^n$ and $\mathbb{R}^m$. Let system f be Lipschitz continuous on $\mathbb{X} \times \mathbb{U}$ with Lipschitz constants L_1^x and L_1^v with respect to the first and the second argument, respectively, and with respect to the metrics induced by V_1 and V_2. Let $L_2^{x_0}$ be the Lipschitz constant of the parametrization with respect to the state x, i.e. the first argument, and let $K_p = 0$. For a given set $\Delta \subset \mathbb{R}^n$ define $k_1 = \sup_{x \in \Delta} V_1(x)$ and the sets*

$$\Omega_j^x := \left\{ x | V_1(x) \leq \left((L_1^x)^j + L_1^v L_2^{x_0} \sum_{r=1}^{j} (L_1^x)^{r-j} \right) k_1 \right\}, \ j = 1, \ldots, N \qquad (6.15)$$

$$\Omega_j^u := \{ u | V_2(u) \leq L_2^{x_0} k_1 \}, \ j = 0, \ldots, N-1. \qquad (6.16)$$

Let $\{z\} \oplus \Delta \subset \mathbb{X}$, and let $\tilde{U}$ such that $\{x_j(z, z, \tilde{U})\} \oplus \Omega_j^x \subset \mathbb{X}$ and $\{u_j(z, z, \tilde{U})\} \oplus \Omega_j^u \subset \mathbb{U}$. Then for all $x \in \{z\} \oplus \Delta$

$$x_j(z, z, \tilde{U}) - x_j(x, x, \tilde{U}) \in \Omega_j^x, \ j = 1, \ldots, N, \qquad (6.17)$$

$$u_j(z, z, \tilde{U}) - u_j(x, x, \tilde{U}) \in \Omega_j^u, \ j = 0, \ldots, N-1 \qquad (6.18)$$

holds. In particular, $x_j(x, x, \tilde{U}) \in \mathbb{X}$ and $u_j(x, x, \tilde{U}) \in \mathbb{U}$ holds.

Proof. As discussed above, the assumptions of the proposition imply that Assumption 6.1 is fulfilled. Relation (6.14) readily yields (6.18) with (6.16). Relation (6.17) with (6.15) is shown recursively for j, $1 \leq j \leq N$. For the initial step $j = 1$, the claim directly follows from inserting (6.14) into (6.13). Let (6.17) be valid for j, $1 \leq j < N$. Again taking (6.13) and plugging in (6.17) and (6.14) for j yields the claim for $j + 1$. $\qquad \square$

The size of the initial set Δ affects the size of all consecutive sets in a linear fashion. Decreasing the size of Δ decreases k_1 and thereby the size of the required sets Ω_j^x and Ω_j^u. In contrast, the influence of the Lipschitz constant L_1^x is exponential. So having a Lipschitz constant $L_1^x > 1$, the sets Ω_j^x quickly become very large which can restrict applicability of the results.

Tube cross sections for incrementally stable systems

A second special case is derived from Assumption 6.1 requiring therein $L_1^x \leq 1$. From a system theoretic perspective, this means that system $\tilde{f}$ is incrementally stable, see (Bayer et al., 2016). In addition we require here that the v_j inputs applied to the trajectory originating from x equal the v_j inputs applied to the test state trajectory, i.e., $p(\tilde{x}, \tilde{U}) = p(z, \tilde{U})$. Clearly, this is achieved via the choice $\tilde{x} = z$, i.e., evaluating the parametrization for all states $x \in \{z\} \oplus \Delta$ at z.

Regarding the parametrization, this means that, in fact, the original parametrization is replaced by one which is piecewise constant in the state. The benefit of this assumption together with having $L_1^x \leq 1$ is that the distance of the state sequences remains in the same level set as the initial conditions of the sequences, i.e., the size of the state tube cross sections remains constant over the prediction horizon. As a consequence, also the difference of the applied input transformation $K_p x$ remains in a level set of constant size. The next result readily follows.

Proposition 6.3 (Tube cross sections for incrementally stable system). *Let Assumption 6.1 hold with $L_1^x \leq 1$. For a given set $\Delta \subset \mathbb{R}^n$ define $k_1 = \sup_{x \in \Delta} V_1(x)$ and the sets*

$$\Omega^x := \{x | V_1(x) \leq k_1\} \tag{6.19}$$

$$\Omega^u := \{u | u = K_p \tilde{x}, \ V(\tilde{x}) \leq k_1\}. \tag{6.20}$$

Let $\{z\} \oplus \Delta \subset \mathbb{X}$, and let $\tilde{U}$ such that $\{x_j(z, z, \tilde{U})\} \oplus \Omega^x \subset \mathbb{X}$ and $\{u_j(z, z, \tilde{U})\} \oplus \Omega^u \subset \mathbb{U}$. Then for all $x \in \{z\} \oplus \Delta$

$$x_j(z, z, \tilde{U}) - x_j(x, z, \tilde{U}) \in \Omega^x, \ j = 1, \ldots, N \tag{6.21}$$

$$u_j(z, z, \tilde{U}) - u_j(x, z, \tilde{U}) \in \Omega^u, \ j = 0, \ldots, N-1 \tag{6.22}$$

holds. In particular, $x_j(x, z, \tilde{U}) \in \mathbb{X}$ and $u_j(x, z, \tilde{U}) \in \mathbb{U}$ holds.

Again, the size of the level sets Ω^x and Ω^u is affected linearly by the maximum level k_1 attained in the initial set Δ. The size of the tube cross sections is constant over the prediction horizon which simplifies matters and is typically by far better applicable than having growing tubes as can be the case in the first special case considered.

Based on these results, we can now formulate a procedure to test feasibility of a parametrized optimization problem for states x in a given set $\hat{\mathcal{D}}$. The procedure is as follows:

1. Find states $\{z_1, \ldots, z_{n_t}\}$ and a set Δ such that (6.7) holds.

2. Compute tube cross sections $\Omega_1^x, \ldots, \Omega_N^x$ and $\Omega_0^u, \ldots, \Omega_{N-1}^u$ as required in Lemma 6.1 by using Proposition 6.2 or 6.3.

3. For all $\{z_1, \ldots, z_{n_t}\}$ formulate and evaluate feasibility test problem (6.11) according to Lemma 6.1.

If all feasibility tests are successful, the transformed parametrized optimization problem (6.3) is feasible for all states in the set $\hat{\mathcal{D}}$.

The main advantage of following this robust MPC inspired strategy instead of the robust optimization based approach was that additional system theoretic insights into the problem could be used to find a suitable constraint tightening. Even more important, the problem transformation into inputs in v coordinates could only be applied based on system theoretic insights and can generally largely reduce conservatism and extend applicability.

6.3 Computing feasible parametrizations: The offline algorithm

6.3.1 The general approach

The approach to compute feasible parametrizations in the nonlinear case goes along the same lines as in the linear case. First, a preliminary parametrization is computed by clustering training data, then a postprocessing step is applied to render the parametrization feasible. As will become evident, dependencies among the single steps are more distinct in the nonlinear setting. To avoid complicating matters unnecessarily, it is reasonable to sample the training states from a regular grid. Determining cluster membership online is then easily achieved by assigning each state to the same cluster as the closest training state, i.e., to essentially generalize cluster membership to hypercubes around the training states. It is reasonable to use the training states and these hypercubes for the refinement procedure as well. Thus, we focus on this case. In the refinement procedure, the results on the feasibility test introduced above will be employed and the parametrization is adjusted to approximate solutions of such feasibility test. To account for that, also the training data is computed for a problem which uses the same tightened constraints as the feasibility test problem. Beyond that, if the constraint tightening is determined using a transformed system $\tilde{f} \neq f$, the parametrization approximates the transformed inputs $v = K_p x - u$ and, hence, these inputs have to be used as training data. The optimization which yields the input trajectories $\mathcal{V} = [v_0^\top, \ldots, v_{N-1}^\top]^\top$ for the training data is given by

$$
\begin{aligned}
\min_{v_0, \ldots, v_{N-1}} \quad & \sum_{j=0}^{N-1} \ell(x_j, K_p x_j + v_j) + V_T(x_N) \\
\text{s.t.} \quad & x_{j+1} = \tilde{f}(x_j, v_j), \qquad j = 0, \ldots, N-1 \\
& x_0 = x, \\
& x_j \in \mathbb{X} \ominus \Omega_j^x, \qquad\qquad j = 0, \ldots, N-1 \\
& K_p x_j + v_j \in \mathbb{U} \ominus \Omega_j^u, \quad j = 0, \ldots, N-1 \\
& x_N \in \mathbb{X}_T \ominus \Omega_N^x.
\end{aligned}
\tag{6.23}
$$

Summarizing, the offline procedure in the nonlinear case is as follows.

1. Select a regular grid of training states covering the desired feasible set.

2. Take the set Δ as the hypercube defined by 2^n neighboring grid points centered at the origin.

3. Find tube cross sections Ω_j^x and Ω_j^u which fulfill (6.9) and (6.10). To this end, knowledge about the system to be controlled and possibly an assumption about the final parametrization[2] can be used and the input transforming matrix K_p can be chosen freely. For suitable systems, Proposition 6.2 or 6.3 are applicable in this step.

4. Compute training data via solving (6.23) for each training state. If feasibility of this problem is largely deteriorated over feasibility of the original non-tightened problem (6.3), selection of a finer grid of training states yields a smaller set Δ, resulting in smaller sets Ω_j^x, Ω_j^u and thus improved feasibility.

5. Run the clustering procedure.

6. Run the refinement procedure.

The latter two items are still kept rather unspecific and will be addressed in the next subsections more in detail.

6.3.2 Adapted clustering algorithm

Next, we present a slightly extended clustering algorithm, which is adapted for the application to nonlinear problems. The extensions are motivated by observations made when applying the general tailored subspace clustering algorithm introduced above to nonlinear problems. Recall that when applying the clustering to linear problems, in most cases compactly shaped clusters in state space were observed, see for example Figure 4.2 and 4.3. In contrast, this was not the case when treating nonlinear systems. In many cases the clusters were spatially extended or not connected in state space and often also outliers located separated from other states in the same cluster were observed. This behavior seems to be due to three reasons. One the one hand, scattered clusters can reflect the actual structure of the data which is approximated via the clustering. On the other hand, in the nonlinear case generally parametrizations were used which depend on the state in a rather general way via a general mapping ϕ. This can cause scattered solutions to be (locally) optimal and, finally, can impede convergence of the algorithm to good solutions which one would generally expect to be shaped regularly in state space. Recall, that deteriorated convergence was also observed in the linear case using constant terms in the parametrization.

Whereas the shape of clusters in state space might seem non-relevant, the scattered clusters turned out to be disadvantageous in the refinement procedure. Rendering parametrizations feasible which have degenerated shape and outliers in state space generally proved to be much harder than it was the case for compactly shaped clusters. First of all states far apart from the cluster center in state space were hard to render feasible. Two intuitive reasons can be given for this finding. First, higher order terms in ϕ get a larger effect

---

[2] Computation of the sets for constraint tightening might require information about the final parametrization which is not yet available at this stage. Thus, it might be necessary to start for example assuming a Lipschitz constant and then a posteriori assess if this assumption is met by the final parametrization. If this is not the case, this procedure has to be repeated using relaxed assumptions as e.g. a larger Lipschitz constant.

for larger states and states far away from the cluster center. Thus, if the weight on the higher order terms is suitable on average for a cluster, it will generally be not optimal for states far away from the cluster center but have a large effect for them. Second, even in a non-convex setting states at the edge of the clusters are more likely to be decisive in terms of constraint satisfaction. As a conclusion, it is reasonable to counteract clusters which are very scattered in state space and promote compactly shaped clusters.

In order to do so, we exploit that the chosen clustering approach can be simply extended and add a term to the clustering objective function which fosters compactly shaped clusters in state space. The resulting algorithm executes a tailored subspace clustering as introduced before in this thesis and combines it with a K-means clustering executed in state space.

The combined subspace/K-means clustering algorithm addresses the optimization problem

$$\min_{M_i, N_i, c_i, \tilde{U}_r, \mu_i^r, \ i=1,\ldots,K, \ r=1,\ldots,s} \tilde{J}_C$$
$$\text{s.t. } \mu_i^r \in \{0,1\} \text{ and } \sum_{i=1}^{K} \mu_i^r = 1, \tag{6.24}$$

with

$$\tilde{J}_C = \sum_{i=1}^{K} \sum_{r=1}^{s} \mu_i^r \left((1-\lambda)\|U_r - M_i\tilde{U}_r - N_iy_r\|_2 + \lambda\|x_r - c_i\|_2 \right)$$

for given $\lambda \in [0,1]$, $M_i \in \mathbb{R}^{mN \times q}$, $N_i \in \mathbb{R}^{mN \times b}$, $c_i \in \mathbb{R}^n$ and $\tilde{U}_r \in \mathbb{R}^q$. Both parts of the objective function are weighted against each other via λ such that in the limiting case $\lambda = 0$ the extended subspace clustering is recovered and for $\lambda = 1$ the clustering becomes pure K-means clustering of the training states.

A slightly different interpretation of the term $\lambda\|x_r - c_i\|_2$ is to see it as a regularization term as it is used in some algorithms which approximate data to reduce complexity of the solutions and avoid over fitting the data. Here, it regularizes the clusters in state space and thereby has a beneficial overall influence on the clustering result.

Optimization (6.24) can be solved via an iterative procedure exactly along the same lines as is done in the tailored subspace clustering algorithm introduced in Algorithm 2. To keep the presentation concise, we only present the single building blocks of the algorithm. A convergence result as for Algorithm 2 holds as well. For the *cluster update* in this case the problems

$$\min_{M_i, N_i, \tilde{U}_r} \sum_{\{r|\mu_i^r=1\}} \|U_r - M_i\tilde{U}_r - N_iy_r\|_2 \tag{6.25}$$

and

$$c_i = \frac{1}{\sum_r \mu_i^r} \sum_{\{r|\mu_i^r=1\}} x_r \tag{6.26}$$

have to be solved. The *cluster assignment* in this case is done via solving

$$\min_{\mu_i^r, \tilde{U}_r} \sum_{i=1}^{K} \sum_{r=1}^{s} \mu_i^r \left((1-\lambda)\|U_r - M_i\tilde{U}_r - N_iy_r\|_2 + \lambda\|x_r - c_i\|_2 \right)$$
$$\text{s.t. } \mu_i^r \in \{0,1\} \text{ and } \sum_{i=1}^{K} \mu_i^r = 1. \tag{6.27}$$

In order to compute parametrizations offline, the procedure to compute compactly shaped clusters is applied in Algorithm 3 replacing therein the basic clustering Algorithm 2. In an example below a parametrization will be computed via these algorithms and the effect of λ on the result will be evaluated.

6.3.3 Refinement of the parametrization

When refining a parametrization such that it is feasible for a given set of (infinitely many) states, the challenges are the same as when testing feasibility. One has to find a way to treat infinitely many different states while keeping the computational effort limited. Accordingly, the refinement approach we suggest here is closely related to the feasibility test results formulated above. Based on achieving feasibility subject to tightened constraints for a number of test states, feasibility for sets around the test states is ensured. An additional challenge in a general nonlinear case is that even adjusting a parametrization directly such that it is feasible for a small number of test states is not trivial. Thus, here we suggest an iterative adjustment procedure which only requires to evaluate feasibility tests and to recompute the matrices M_i and N_i in the parametrization based on a weighted least squares/low rank approximation problem. More in detail, the suggested procedure iteratively tests feasibility of the parametrization for test states and improves approximation of a feasible solution for those test states which were found to be infeasible until feasibility for all test states is achieved or a maximum iteration number is reached. Thereby, the algorithm basically distributes "the overall approximation capability" of the parametrization in a systematic fashion such that ideally feasibility of all tightened test problems is achieved.

As the refinement procedure has to be applied for each part p_i of the parametrization independently, we restrict attention to one part p_i in the following. The task is to achieve feasibility of the parametrized problem (6.3) for all states x in a set $\hat{\mathcal{D}}_i \subset \mathbb{R}^n$. Like for the feasibility test, a set $\Delta \subset \mathbb{R}^n$ and test states $\{z_1, \ldots, z_{n_t} \in \mathbb{R}^n\}$ have to be chosen such that

$$\hat{\mathcal{D}}_i \subseteq \Delta \oplus \{z_1, \ldots, z_{n_t}\} \tag{6.28}$$

holds. Using the sets Δ and Proposition 6.2 or 6.3, a tightened feasibility test problem (6.11) can then be formulated based on Lemma 6.1.

The weighted least squares/low rank approximation problem we employ is a weighted version of (3.10) (or equivalently (6.25)) given by

$$\min_{M_i, N_i, \tilde{U}_r} \sum_{r=1}^{n_t} w_r \|U_r - M_i \tilde{U}_r - N_i y_r\|_2 \tag{6.29}$$

wherein $w_1, \ldots, w_{n_t} > 0$ are the weights, U_r is a solution of (6.23) for test state z_r and $y_r = \phi(z_r)$ holds. This problem can be solved very efficiently as discussed for (3.10).

The complete refinement procedure is formalized in Algorithm 15. Note that feasibility of each part p_i for $i = 1, \ldots, K$ has to be achieved separately via executing the algorithm. If therein in line 6 feasibility for all test states is detected, Lemma 6.1 allows to directly conclude on overall feasibility assuming that the ingredients to the algorithm have been chosen properly. A straight forward choice for the test points in the algorithm is to use the training states from clustering which were assigned to the respective cluster. In order to be able to apply a less restrictive tightening, an increased number of test points can be used.

Algorithm 15 Refinement of parametrization for nonlinear systems

---

Input: test points z_r and $y_r = \phi(z_r)$, $r = 1, \ldots, n_t$, maximum iterations $i_{\max} \in \mathbb{N}$, constant $\eta > 1$

1: for each z_r compute U_r as minimizer of (6.23)
2: initialize $it = 0$ and $w_r = 1$ for $r = 1, \ldots, n_t$
3: **while** $it < i_{\max}$ **do**
4:     determine M_i, N_i via (6.29)
5:     **for** $r = 1, \ldots, n_t$ **do**
6:         Check feasibility of (6.11) for $z = z_r$, if infeasible set $w_r = \eta w_r$
7:     **end for**
8:     **if** any w_r has changed w.r.t. previous iteration **then**
9:         set $it = it + 1$ // param. not feasible, further iteration needed
10:    **else**
11:        quit. // feasible solution found
12:    **end if**
13: **end while**
Output: M_i, N_i

The main advantage of this approach is its simplicity and scalability. All feasibility tests (6.11) can be executed independently in a sequential or parallel manner. This is well tractable as each test has a complexity comparable to the resulting online optimization problem. The weighted M_i, N_i update (6.29) which incorporates all test states at once is completely independent of the underlying nonlinear problem and is thus not affected by any possibly obstructive properties thereof.

6.4 Example and evaluation

Next we present a simple numerical example to illustrate the methods introduced and then evaluate and discuss the results of this chapter.

6.4.1 Numerical example

The class of systems considered in this chapter is too general to cover via numerical examples all relevant aspects and situations which can arise when applying the proposed methods. We restrict ourselves to the presentation of one illustrative numerical example to show how and that the proposed methods are applicable and illustrate some very general trends. Clearly, this is by no means exhaustive nor generally representative.

Example 6.4. *This example has been treated before in (Bayer et al., 2016; Limon Marruedo et al., 2002). The system*

$$x^+ = f(x, u) = \begin{bmatrix} 0.55x_a + 0.12x_b + (0.01 - 0.6x_a + x_b)\,u \\ 0.67x_b + (0.15 + x_a - 0.8x_b)\,u \end{bmatrix} \tag{6.30}$$

with state $x = [x_a, x_b]^\top$ is considered where the state x is constrained to the set $\mathbb{X} = [-5, 5]^2$.

and the input u is constrained[3] to $\mathbb{U} = [-0.1, 0.1]$. We use the stage cost $\ell(x, u) = 0.5\|x\|^2 + 5u^2$. The terminal control law $\kappa(x) = K_p x$, with $K_p = -[0.002, \ 0.019]$, the terminal cost function $V_T(x) = x^\top P x$ and the terminal constraint set $\mathbb{X}_T = \{x \in \mathbb{R}^2 | 0.5 x^\top P x \leq 1.124\}$ with $P = \left[\begin{smallmatrix} 2.867 & 0.298 \\ 0.298 & 3.777 \end{smallmatrix}\right]$ are used (Bayer et al., 2016). They can be determined based on (Rawlings and Mayne, 2009, Chapter 2.5). A prediction horizon of $N = 10$ is used.

In the literature it was shown that the transformed system $\tilde{f}(x, v) = f(x, u + K_p x)$ is incrementally stable, in particular $\tilde{f}$ fulfills the assumptions of Proposition 6.3 with $V_1(x) = \|x\|_1$. Therefore, Proposition 6.3 is applied to compute the constraint tightening required for the refinement of the parametrization and we stick to the notation of Proposition 6.3 in the following.

We use a regular grid of training data for the clustering and also for the refinement of the parametrizations. The distance of two neighboring grid points along a coordinate axis is chosen to be equal in both coordinate axis directions and will be denoted by δ. The sets Δ used in the refinement and in the feasibility tests are then defined by four neighboring grid points and are given by $\Delta = [-\frac{1}{2}\delta, \frac{1}{2}\delta]^2$. It results that the constant k_1 defined in Proposition 6.3 is given by $k_1 = \delta$ and thus $\Omega^x = \{x | \ \|x\|_1 \leq \delta\}$. Here it is convenient to use the box $\overline{\Omega}^x = [-\delta, \delta]^2$ which is a superset of Ω^x to tighten the state and the terminal constraints. Correspondingly, the set Ω^u is contained in the set $\overline{\Omega}^u = \{u | K_p x, \ x \in \overline{\Omega}^x\}$ which will be used to tighten the input constraints. In order to tighten the elliptic terminal constraint, we first compute a polytopic inner approximation thereof and then apply the tightening to this set.

Due to the decision to use the same states for clustering and for refinement, the number of grid points used directly influences the required constraint tightening for the refinement. The example nicely illustrates how the number of grid points influences $\overline{\Omega}^x$ and $\overline{\Omega}^u$ via δ. A direct consequence is that a minimum number of grid points is required to ensure that the tightened constraints define non-empty sets and even more grid points might be necessary to improve feasibility of the tightened problems.

Using a grid of 13×13 training states which exactly covers $\mathbb{X}$ results in an empty tightened terminal constraint set and using a grid of 14×14 states the tightened terminal constraint set is non-empty but the overall set of feasible states of problem (6.23) is considerably smaller than for the corresponding non-tightened problem. We use a grid of 16×16 training states which further relaxes the required constraint tightening such that the tightened problem is feasible for a large set of states.

We next fixed the hyperparameters $q = 1$ and $\phi(x) = [x^\top, x_a^2, x_b^2, x_a^3, x_b^3, x_a^4, x_b^4, x_a x_b]^\top$ and decided to run the clustering from a rather large number of 100 initial conditions in order to reduce influence of the convergence behavior of the clustering. We tested for several values of K different values for λ and found that $\lambda = 0.004$ yields clusters of reasonably compact shape in state space. In Figure 6.1 the clusters in state space obtained for $K = 8$ and using $\lambda = 0$ (left hand side) and $\lambda = 0.004$ (right hand side) are compared. Obviously, clusters on the left hand side are very scattered whereas they are compactly shaped on the right hand side.

An interesting observation could be made comparing the clustering results in terms of J_C as defined in (3.9), i.e. measuring only approximation accuracy of the input sequences. For $K \leq 5$, the achieved J_C was lower using $\lambda = 0$ than using $\lambda = 0.004$. This is what

---

[3] Note that with respect to (Bayer et al., 2016) relaxed input constraints are used as this results in more diverse input sequences which are more challenging to parametrize yielding more insightful results.

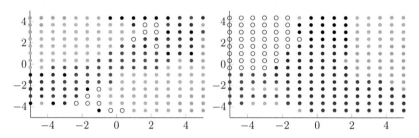

Figure 6.1: Clusters in state space using $\lambda = 0$ (left) and using $\lambda = 0.004$ (right).

one would expect as $\lambda = 0$ means to exactly address minimization of J_C whereas $\lambda = 0.004$ means to minimize $\tilde{J}_C \neq J_C$. The findings using $K \geq 6$ were in contrast to that. In these cases, in fact lower values of J_C were observed using $\lambda = 0.004$ instead of using $\lambda = 0$. This indicates that for larger numbers of clusters using $\lambda = 0.004$ actually improves convergence of the clustering to good solutions despite the fact that it addresses a slightly different objective.

For the different parametrizations the refinement (Algorithm 15) was tested using therein $\eta = 2$ and a maximum of $i_{max} = 20$ iterations. For $\lambda = 0.004$ the lowest value of K such that the refinement was feasible was $K = 8$ (c.f. Figure 6.1). In this case, two clusters were immediately feasible, the remaining ones required in between 6 and 13 iterations to achieve feasibility. Using $\lambda = 0$ and $K = 8$ the parametrizations could not be rendered feasible.

The complete offline computation times for this example using $K = 8$ and $\lambda = 0.004$ are 41 s to generate the training data, 17 s for the clustering and 9 s for the refinement.

6.4.2 Evaluation

Let us next briefly evaluate some general properties of the proposed results. A main characteristics in the nonlinear case is that there is a large number of decisions to make in the algorithm and dependencies and effects thereof are rather involved. We consider properties of the online algorithm and then trace back the decisions in the offline algorithm which influence them. In the online algorithm, generally speaking, at each time step Problem (6.4) is solved for the current state x and the result is used to obtain an optimized input sequence for the current state x. In detail, this requires the following steps:

1. Obtain current state x

2. Find the right part p_i of the parametrization p for the current state x.

3. If $\tilde{x} \neq x$ is to be used in (6.4), find the right $\tilde{x}$ for the current state x.

4. Solve (6.4) for x, the chosen p_i and $\tilde{x}$.

5. If optimization is feasible and solution yields sufficiently low predicted open-loop costs:
   Compute the resulting input sequence $p_i(\tilde{x}, \tilde{U}^*)$.
   Else
   Compute candidate solution from previous solution.

6. If $K_p \neq 0$, transform first element v of chosen input sequence into input u.

7. Apply first element u of chosen (transformed) input sequence to the plant and return to step 1.

In step 2, typically the closest training state is searched and its cluster membership is used. Thus, the complexity and the required amount of data to store for this step is mainly determined by the amount of training data used. In step 3, $\tilde{x} \neq x$ has to be used if in the refinement the parametrization was evaluated for the test state (as was done in the results for incrementally stable systems) rather than bounding the effect of using a state close to the test state (as was done in the results for Lipschitz continuous systems). In this case, $\tilde{x}$ has to be chosen equal to the closest test state used in the refinement.[4] Thus, the complexity and the required amount of data for this online step is mainly determined by the number of test states used in the refinement offline. On the other hand, using more test states in the offline refinement procedure reduces the required amount of constraint tightening and thereby reduces conservatism and improves feasibility. Such situation can be avoided completely using $\tilde{x} = x$ (as it was suggested for Lipschitz continuous systems). This makes offline calculations more challenging but simplifies the online procedure and makes it independent of the test states used in the refinement. Step 5 in the online procedure becomes necessary if a system $\tilde{f} \neq f$ is used as in this case $p_i(\tilde{x}, \tilde{U}^*)$ is not an admissible input sequence for the original problem. As mentioned above, such transformation could in principle be executed offline, yielding a transformed parametrization.

Generally, the results of this chapter are meant to be seen as a "proof of concept", to illustrate that the general method is applicable to nonlinear problems as well, and to introduce some ideas which can be useful in this setting. The Lipschitz continuity based results are in principle applicable to a rather large class of systems. Yet, application of these results quickly becomes very conservative if the Lipschitz constant of the system dynamics is larger than 1 due to its exponential influence. In this case a global consideration of the state space might not be the best approach and subdividing the state space or the prediction horizon into several parts might improve applicability. If an incrementally stable system (or a system which can be rendered incrementally stable via a suitable input transformation) is to be controlled, the corresponding results are very well applicable as the tube cross sections obtained for constraint tightening have constant size along the prediction horizon. On the other hand, the class of problems covered by these results is rather restricted. In order to apply the semi-explicit MPC approach to a given actual nonlinear control problem tailoring the results by combining ideas of these approaches or applying any of the referenced methods to bound the reachable states will generally improved the results largely.

---

[4] An underlying assumption here is that the sets Δ in the refinement have been chosen such that all states x assigned to test state $\tilde{x}$ are contained in $\tilde{x} \oplus \Delta$ and, similarly, that the cluster membership used in the refinement is compatible with the decision made in step 2.

6.5 Summary and extensions

6.5.1 Summary

In this chapter we addressed semi-explicit MPC for nonlinear systems and in particular presented results regarding the offline part of the approach. A subspace clustering algorithm which allows to promote clusters which are compactly shaped in state space was introduced. Application of this algorithm turned out beneficial for nonlinear problems. A result to test feasibility of the parametrized optimization problem for a given set of states based on a finite number of feasibility test problems was proposed. The result was based on a general assumption which was then shown to be covered by the classes of Lipschitz continuous systems as well as incrementally stable systems. The adapted clustering algorithm and the feasibility test were extended and integrated into a complete offline algorithm which, if executed successfully, yields parametrizations which are guaranteed to result in a feasible parametrized optimization problem for a given set of states. The results were illustrated via a numerical example where the method could be applied maintaining all theoretical guarantees.

6.5.2 Extensions

The results of this chapter provide a starting point for various extensions and adaptations of the general principle applied. In the refinement procedure, constraint tightening could be done based on various other approaches to bound the reachable states. Besides addressing feasibility only in the refinement, the procedure could also be used to address open-loop control performance at the test states by extending the feasibility test problems to ones with suitable objective functions and taking suboptimality observed therein into account.

For a given problem, tailoring the employed methods to compute reachable states would certainly render the approach less conservative and more powerful. This could exploit certain properties of the system, either regarding exact properties the system possesses or exploiting further structural system properties. The latter approach could for example go along the lines of the results found in (Rakovic et al., 2006) for robust MPC.

Furthermore, for specific system classes feasibility could be guaranteed independently of the clustering based parametrization itself, but based on an additional basis vector consisting of a feasible solution to the optimization problem. This is conceptually similar to the third approach to find feasible parametrizations for linear systems introduced in Subsection 4.3.3 of this thesis and could, e.g., go along the lines followed in (Bloemen et al., 2002).

Moreover, it might be worth investigating possible structural simplifications of the online optimization in the case of using scalar decision variables ($q = 1$). Beyond that, extending the results to robust MPC problems seems a natural step to take as the techniques typically employed in robust MPC are already used for the current results.

Chapter 7

Conclusion

In the following, the main results of this thesis are summarized and possible directions for future research are indicated.

7.1 Summary

Driven by the desire to open up new application areas for MPC, a major recent and ongoing research effort addresses the development of fast and computationally efficient MPC schemes. In this thesis, we propose the concept of a very generally applicable, novel semi-explicit MPC scheme and thereby make a contribution towards this endeavor. The semi-explicit approach consists of a first offline phase during which state-dependent parametrizations for optimal input sequences are computed based on training data. In a second online phase, the parametrizations are applied in an MPC scheme with reduced computational load. This way, the scheme combines aspects and advantages of online optimization based MPC and of explicit MPC. We contribute in the thesis various algorithms required for the scheme and establish system theoretic properties of the scheme.

In Chapter 3 we introduced the general semi-explicit MPC scheme, presented its building blocks and evaluated problem-independent properties of the scheme. In particular, a tailored subspace clustering algorithm, a way to compute the parametrizations based on the subspace clustering algorithm and the online semi-explicit MPC scheme were presented. Guarantees on system theoretic properties as strong and recursive feasibility, closed-loop stability and a bound on closed-loop control performance were established.

Chapter 4 was dedicated to semi-explicit MPC for linear systems. Results exploiting properties of this problem class were presented including readily applicable semi-explicit MPC algorithms for linear systems. Specifically, three methods to offline compute parametrizations with guaranteed feasibility were proposed, an offline stability test which allows to employ a simplified asymptotically stabilizing online MPC scheme was formulated and a bound on the relative closed-loop cost increase was proven.

In Chapter 5 we proposed an extremely simple online solution strategy for the special case of using a univariate parametrization in linear MPC problems. A basic algorithm, two simplifications thereof as well as a detailed complexity evaluation of the approach were presented. Superiority of the proposed strategy to explicit solution strategies was shown.

In Chapter 6 we established results which allow to apply the general method to specific nonlinear MPC problems. Specifically, an algorithm to influence the shape of clusters in state space was presented and a strategy to test feasibility of parametrized MPC problems for a given set of states was formulated and it was turned into an algorithm to compute parametrizations with guaranteed feasibility. These general results were elaborated specifically for the classes of Lipschitz continuous and incrementally stable systems.

A central aspect within the thesis was the formulation of algorithms to compute parametrizations which result in parametrized MPC problems with guaranteed feasibility for a given set of states. Throughout the thesis five different approaches were employed to this end. The first approach extensively used training data and basically enlarged and refined obtained clusters in state space, the second approach explicitly included the desired feasible set into the clustering, the third approach simply exploited feasibility of an added independent part, the fourth approach mainly followed a robust optimization strategy and available Lipschitz constants, whereas the fifth approach employed tube-based robust MPC techniques. In summary, these different approaches are applicable to a range of different problem classes and, hence, make the semi-explicit MPC approach applicable to them. Throughout the thesis several numerical examples were presented, illustrating applicability, properties and benefits of the proposed semi-explicit MPC scheme. In particular, it was shown that the proposed method can have advantages over existing methods and thereby extends applicability of MPC.

7.2 Outlook

Semi-explicit MPC as proposed in this thesis combines and employs in a novel way a number of concepts yielding a complete MPC scheme. Even though readily applicable algorithms and MPC schemes were proposed, the presented results remain to some extent exemplary for the general semi-explicit idea. Much room for refinements and smaller extensions of the presented results is left but, even more, a starting point for numerous directions of future research carrying on this general idea much further is provided.

Immediate smaller extensions were indicated and discussed in the respective previous sections of this thesis. They comprise aspects as finding further approaches to compute feasible parametrizations, addressing and exploiting known structure in mpP solutions, selecting training data more elaborately, application of the methods with slightly different focus as reduction of the number of constraints or recognition of structure in mpP solutions as well as simplifying the whole approach by omitting strict feasibility and constraint satisfaction guarantees but instead e.g. softening constraints.

Some more fundamental extensions are indicated in the following. Application and tailoring of the semi-explicit MPC scheme for system classes beyond what has been addressed in this thesis is possible and could be considered in future research. On the one hand, extended system classes as hybrid or further nonlinear system classes could be covered via adaptations of the presented algorithms. On the other hand, given a specific (practical) control problem, exploiting more exhaustively properties of the system and MPC problem within the semi-explicit scheme would extend applicability and reduce conservatism thereof.

Similarly, application of the general approach in a wide range of MPC schemes is in principle possible and could be investigated more in detail in future research. Robust MPC schemes are certainly amenable to the approach and the main question here seems to be if the computational simplifications achieved still justify the overall complexity the resulting procedure could possibly have. Application of the semi-explicit concept in unconstrained and economic MPC schemes would certainly be beneficial. Relevant issues in this case are system theoretic properties which would have to be established when parametrizations are used in these schemes.

More innovative extensions and descendants of the semi-explicit MPC scheme are certainly obtained originating from a more abstract interpretation of its principle: From an abstract point of view, the scheme generates data in form of optimal input sequences, applies a data mining method to reveal and extract only relevant structural properties of the data and then exploits this structure for an online simplification. This is in contrast to existing approaches which typically use all information or no information at all.

The clustered quantities do not necessarily have to be continuous-valued and they do not have to be (input) signals. On the one hand, an obvious direction is to tailor the semi-explicit MPC approach to systems with discrete-valued input signals and use therein an adapted clustering algorithm to find approximations of such signals which can be employed beneficially online. A more interesting direction, which is worth being investigated, is to consider the information whether or not the single constraints are activity for different states as a quantity of interest. For each training state activity of constraints at the optimal solution could be represented by a Boolean vector. Clustering of these vectors might enable a compact representation of useful information exploitable online to simplify computations. A slightly related strategy has been pursued in (Zhu et al., 2015).

Furthermore, giving up some beneficial structural properties of the parametrizations, more general parametrizations could be sought and applied in a very similar framework as proposed in this thesis. To this end, the general "trick" to map quantities into a higher-dimensional space where linear methods are applicable and are more powerful might be adaptable. Establishing details on the procedure as well as investigating the effect of using more complex mappings from the parameter to the predicted input sequences on numerical online complexity are necessary.

Still loosely related to the presented work, it would be interesting to investigate ways to reveal more general properties and patterns of (input) signals such as jumps or sudden changes in the behavior and to find ways to translate these properties into useful parametrizations or, more generally, into some information beneficial for the online optimization.

Bibliography

P. K. Agarwal and N. H. Mustafa. K-means projective clustering. In *Proceedings of the 23rd ACM SIGMOD-SIGACT-SIGART symposium on principles of database systems*, pages 155–165. ACM, 2004.

R. P. Aguilera and D. E. Quevedo. On the stability of MPC with a finite input alphabet. *IFAC Proceedings Volumes*, 44(1):7975–7980, 2011.

M. Alamir. A framework for real-time implementation of low-dimensional parameterized NMPC. *Automatica*, 48(1):198–204, 2012.

M. Alamir, A. Murilo, R. Amari, P. Tona, R. Fürhapter, and P. Ortner. On the use of parameterized NMPC in real-time automotive control. In *Automotive model predictive control*, pages 139–149. Springer, 2010.

T. Alamo, J. M. Bravo, M. J. Redondo, and E. F. Camacho. A set-membership state estimation algorithm based on DC programming. *Automatica*, 44(1):216–224, 2008.

W. Alber. Extensions of a clustering based MPC parametrization algorithm. Master's thesis, Institute for Systems Theory and Automatic Control (IST), University of Stuttgart, Germany, 2013.

A. Alessio and A. Bemporad. A survey on explicit model predictive control. In *Nonlinear Model Predictive Control: Towards New Challenging Applications*, pages 345–369. Springer, 2009.

A. Aswani, H. Gonzalez, S. S. Sastry, and C. Tomlin. Provably safe and robust learning-based model predictive control. *Automatica*, 49(5):1216–1226, 2013.

D. Axehill. Controlling the level of sparsity in MPC. *Systems & Control Letters*, 76:1–7, 2015.

B. Bank, J. Guddat, D. Klatte, B. Kummer, and Tammer. *Non-linear parametric optimization*. Birkhäuser, 1983.

F. A. Bayer, F. D. Brunner, M. Lazar, M. Wijnand, and F. Allgöwer. A tube-based approach to nonlinear explicit MPC. In *Proceedings of the 55th IEEE Conference on Decision and Control, 2016*, pages 4059–4064. IEEE, 2016.

A. Bemporad, M. Morari, V. Dua, and E. N. Pistikopoulos. The explicit linear quadratic regulator for constrained systems. *Automatica*, 38(1):3–20, 2002.

P. Biswas, P. Grieder, J. Löfberg, and M. Morari. A survey on stability analysis of discrete-time piecewise affine systems. *IFAC Proceedings Volumes*, 38(1):283–294, 2005.

H. Bloemen, M. Cannon, and B. Kouvaritakis. An interpolation strategy for discrete-time bilinear MPC problems. *IEEE Transactions on Automatic Control*, 47(5):775–778, 2002.

F. Borrelli, M. Baotić, J. Pekar, and G. Stewart. On the computation of linear model predictive control laws. *Automatica*, 46(6):1035–1041, 2010.

F. Borrelli, A. Bemporad, and M. Morari. Predictive control for linear and hybrid systems. available online at: http://www.mpc.berkeley.edu/mpc-course-material, 2015.

S. Boyd and L. Vandenberghe. *Convex Optimization*. Cambridge University Press, 2009.

P. S. Bradley and U. M. Fayyad. Refining initial points for K-means clustering. In *ICML*, volume 98, pages 91–99. Citeseer, 1998.

P. S. Bradley and O. L. Mangasarian. k-plane clustering. *Journal of Global Optimization*, 16(1):23–32, 2000.

J. M. Bravo, T. Alamo, and E. F. Camacho. Robust MPC of constrained discrete-time nonlinear systems based on approximated reachable sets. *Automatica*, 42(10):1745–1751, 2006.

R. Cagienard, P. Grieder, E. Kerrigan, and M. Morari. Move blocking strategies in receding horizon control. *Journal of Process Control*, 17(6):563–570, 2007.

M. Cannon. Efficient nonlinear model predictive control algorithms. *Annual Reviews in Control*, 28(2):229–237, 2004.

A. Chakrabarty, V. Dinh, M. J. Corless, A. E. Rundell, S. H. Żak, and G. T. Buzzard. Support vector machine informed explicit nonlinear model predictive control using low-discrepancy sequences. *IEEE Transactions on Automatic Control*, 62(1):135–148, 2017.

H. Chen and F. Allgöwer. A quasi-infinite horizon nonlinear model predictive control scheme with guaranteed stability. *Automatica*, 34(10):1205–1217, 1998.

L. Chisci, A. Lombardi, and E. Mosca. Dual-receding horizon control of constrained discrete time systems. *European Journal of Control*, 2(4):278–285, 1996.

C. Danielson and F. Borrelli. Symmetric linear model predictive control. *IEEE Transactions on Automatic Control*, 60(5):1244–1259, 2015.

M. S. Darup and M. Mönnigmann. Low complexity suboptimal explicit NMPC. *IFAC Proceedings Volumes*, 45(17):406–411, 2012.

D. M. de La Pena, T. Alamo, D. R. Ramirez, and E. F. Camacho. Min–max model predictive control as a quadratic program. *IET Control Theory & Applications*, 1(1): 328–333, 2007.

L. Del Re, F. Allgöwer, L. Glielmo, C. Guardiola, and I. Kolmanovsky. *Automotive model predictive control: models, methods and applications*. Springer, 2010.

S. Di Cairano. An industry perspective on MPC in large volumes applications: Potential benefits and open challenges. In *Proceedings of the 4th IFAC Conference on Nonlinear Model Predictive Control*, pages 52–59. Elsevier, 2012.

M. Diehl, H. J. Ferreau, and N. Haverbeke. Efficient numerical methods for nonlinear MPC and moving horizon estimation. In *Nonlinear model predictive control*, pages 391–417. Springer, 2009.

A. Domahidi, M. N. Zeilinger, M. Morari, and C. N. Jones. Learning a feasible and stabilizing explicit model predictive control law by robust optimization. In *Proceedings of the 50th IEEE Conference on Decision and Control and European Control Conference*, pages 513–519, Orlando, FL, USA, 2011.

A. Domahidi, A. U. Zgraggen, M. N. Zeilinger, M. Morari, and C. N. Jones. Efficient interior point methods for multistage problems arising in receding horizon control. In *Proceedings of the 51st IEEE Conference on Decision and Control*, pages 668–674. IEEE, 2012.

K. Duan, S. S. Keerthi, and A. N. Poo. Evaluation of simple performance measures for tuning SVM hyperparameters. *Neurocomputing*, 51:41–59, 2003.

E. Elhamifar and R. Vidal. Sparse subspace clustering: Algorithm, theory, and applications. *IEEE transactions on pattern analysis and machine intelligence*, 35(11):2765–2781, 2013.

M. Ellis, H. Durand, and P. D. Christofides. A tutorial review of economic model predictive control methods. *Journal of Process Control*, 24(8):1156–1178, 2014.

N. P. Faísca, V. Dua, B. Rustem, P. M. Saraiva, and E. N. Pistikopoulos. Parametric global optimisation for bilevel programming. *Journal of Global Optimization*, 38(4):609–623, 2007.

H. J. Ferreau, H. G. Bock, and M. Diehl. An online active set strategy to overcome the limitations of explicit MPC. *International Journal of Robust and Nonlinear Control*, 18 (8):816–830, 2008.

A. V. Fiacco. *Introduction to sensitivity and stability analysis in nonlinear programming*. Academic Press, Inc., 1984.

K. R. Gabriel. Least squares approximation of matrices by additive and multiplicative models. *Journal of the Royal Statistical Society. Series B (Methodological)*, pages 186–196, 1978.

E. G. Gilbert and K. T. Tan. Linear systems with state and control constraints: The theory and application of maximal output admissible sets. *IEEE Transactions on Automatic control*, 36(9):1008–1020, 1991.

G. Goebel and F. Allgöwer. Obtaining and employing state dependent parametrizations of prespecified complexity in constrained MPC. In *Proceedings of the 52nd IEEE Conference on Decision and Control*, pages 7077–7082, Florence, Italy, 2013.

G. Goebel and F. Allgöwer. Improved state dependent parametrizations including a piecewise linear feedback for constrained linear MPC. In *Proceedings of the 2014 American Control Conference*, pages 4192–4197, Portland, OR, USA, 2014a.

G. Goebel and F. Allgöwer. State-dependent parametrizations for nonlinear MPC. In *Proceedings of the 19th IFAC World Congress*, pages 1005 – 1010. IFAC, 2014b.

G. Goebel and F. Allgöwer. A simple semi-explicit MPC algorithm. In *Proceedings of the 5th IFAC Conference on Nonlinear Model Predictive Control*, pages 490–495. Elsevier, 2015.

G. Goebel and F. Allgöwer. New results on semi-explicit and almost explicit MPC algorithms. *at-Automatisierungstechnik*, 65(4):245–259, 2017a.

G. Goebel and F. Allgöwer. Semi-explicit MPC based on subspace clustering. *Automatica*, 83:309–316, 2017b.

R. Gondhalekar and J. Imura. Least-restrictive move-blocking model predictive control. *Automatica*, 46(7):1234–1240, 2010.

A. Grancharova and T. A. Johansen. Semi-explicit distributed NMPC. In *Explicit Nonlinear Model Predictive Control*, pages 209–231. Springer, 2012.

L. Grüne. NMPC without terminal constraints. *IFAC Proceedings Volumes*, 45(17):1–13, 2012.

L. Grüne and J. Pannek. *Nonlinear model predictive control. Theory and Algorithms.* Springer, 2011.

J. Guddat, F. G. Vazquez, and H. T. Jongen. *Parametric optimization: singularities, pathfollowing and jumps.* Springer, 1990.

I. Gurobi Optimization. Gurobi optimizer reference manual, 2016. URL http://www.gurobi.com.

N. Hara and A. Kojima. Reduced order model predictive control for constrained discrete-time linear systems. *International Journal of Robust and Nonlinear Control*, 22(2):144–169, 2012.

M. Herceg, M. Kvasnica, C. N. Jones, and M. Morari. Multi-Parametric Toolbox 3.0. In *Proceedings of the 2013 European Control Conference*, pages 502–510, Zürich, Switzerland, 2013.

T. Hofmann, B. Schölkopf, and A. J. Smola. Kernel methods in machine learning. *The annals of statistics*, pages 1171–1220, 2008.

W. W. Hogan. Point-to-set maps in mathematical programming. *Siam Review*, 15(3):591–603, 1973.

D. Hrovat, S. Di Cairano, H. E. Tseng, and I. V. Kolmanovsky. The development of model predictive control in automotive industry: A survey. In *Proceedings of the 2012 IEEE International Conference on Control Applications*, pages 295–302. IEEE, 2012.

C.-W. Hsu and C.-J. Lin. A comparison of methods for multiclass support vector machines. *IEEE Transactions on Neural Networks*, 13(2):415–425, 2002.

J. L. Jerez, E. C. Kerrigan, and G. A. Constantinides. A sparse and condensed QP formulation for predictive control of LTI systems. *Automatica*, 48(5):999–1002, 2012.

J. L. Jerez, P. J. Goulart, S. Richter, G. A. Constantinides, E. C. Kerrigan, and M. Morari. Embedded online optimization for model predictive control at megahertz rates. *IEEE Transactions on Automatic Control*, 59(12):3238–3251, 2014.

T. A. Johansen. Approximate explicit receding horizon control of constrained nonlinear systems. *Automatica*, 40(2):293–300, 2004.

C. N. Jones and E. Kerrigan. Predictive control for embedded systems. *Optimal Control Applications and Methods*, 36(5):583–584, 2015.

C. N. Jones and M. Morari. Approximate explicit MPC using bilevel optimization. In *Proceedings of the 2009 European Control Conference*, pages 2396–2401, Budapest, Hungary, 2009.

C. N. Jones and M. Morari. Polytopic approximation of explicit model predictive controllers. *IEEE Transactions on Automatic Control*, 55(11):2542–2553, 2010.

A. Joos, P. Heritier, C. Huber, and W. Fichter. Method for parallel FPGA implementation of nonlinear model predictive control. *IFAC Proceedings Volumes*, 45(1):73–78, 2012.

M. Jost and M. Mönnigmann. Accelerating model predictive control by online constraint removal. In *Proceedings of the 52nd IEEE Conference on Decision and Control*, pages 5764–5769. IEEE, 2013a.

M. Jost and M. Mönnigmann. Accelerating online MPC with partial explicit information and linear storage complexity in the number of constraints. In *Proceedings of the 2013 European Control Conference*, pages 35–40. IEEE, 2013b.

M. Jost, G. Pannocchia, and M. Mönnigmann. Online constraint removal: Accelerating MPC with a Lyapunov function. *Automatica*, 57:164–169, 2015.

M. Jost, G. Pannocchia, and M. Mönnigmann. Accelerating linear model predictive control by constraint removal. *European Journal of Control*, 35:42–49, 2017.

Y. Kang and J. K. Hedrick. Linear tracking for a fixed-wing UAV using nonlinear model predictive control. *IEEE Transactions on Control Systems Technology*, 17(5):1202–1210, 2009.

S. Keerthi and K. Sridharan. Solution of parametrized linear inequalities by fourier elimination and its applications. *Journal of Optimization Theory and Applications*, 65 (1):161–169, 1990.

E. C. Kerrigan. *Robust constraint satisfaction: Invariant sets and predictive control*. PhD thesis, University of Cambridge, 2001.

137

B. Khan, G. Valencia-Palomo, J. Rossiter, C. Jones, and R. Gondhalekar. Long horizon input parameterisations to enlarge the region of attraction of MPC. *Optimal Control Applications and Methods*, 37(1):139–153, 2014.

B. Kouvaritakis and M. Cannon. *Model Predictive Control: Classical, Robust and Stochastic*. Springer, 2015.

M. Kvasnica, P. Grieder, M. Baotić, and M. Morari. Multi-parametric toolbox (MPT). *Hybrid systems: computation and control*, pages 121–124, 2004.

M. Kvasnica, J. Löfberg, and M. Fikar. Stabilizing polynomial approximation of explicit MPC. *Automatica*, 47(10):2292–2297, 2011.

M. Kvasnica, J. Hledík, I. Rauová, and M. Fikar. Complexity reduction of explicit model predictive control via separation. *Automatica*, 49(6):1776–1781, 2013.

M. Kvasnica, B. Takács, J. Holaza, and S. Di Cairano. On region-free explicit model predictive control. In *Proceedings of the 54th IEEE Conference on Decision and Control, 2015*, pages 3669–3674. IEEE, 2015.

J. H. Lee. Model predictive control: Review of the three decades of development. *International Journal of Control, Automation and Systems*, 9(3):415, 2011.

D. Li, Y. Xi, and Z. Lin. An improved design of aggregation-based model predictive control. *Systems & Control Letters*, 62(11):1082–1089, 2013.

D. Limon, J. Bravo, T. Alamo, and E. Camacho. Robust MPC of constrained nonlinear systems based on interval arithmetic. *IEE Proceedings-Control Theory and Applications*, 152(3):325–332, 2005.

D. Limon Marruedo, T. Alamo, and E. Camacho. Input-to-state stable MPC for constrained discrete-time nonlinear systems with bounded additive uncertainties. In *Proceedings of the 41st IEEE Conference on Decision and Control, 2002*, volume 4, pages 4619–4624. IEEE, 2002.

C. Liu, W.-H. Chen, and J. Andrews. Piecewise constant model predictive control for autonomous helicopters. *Robotics and Autonomous Systems*, 59(7):571–579, 2011.

J. Löfberg. YALMIP : A toolbox for modeling and optimization in MATLAB. In *Proceedings of the CACSD Conference*, Taipei, Taiwan, 2004. URL http://users.isy.liu.se/johanl/yalmip.

S. Longo, E. C. Kerrigan, K. V. Ling, and G. A. Constantinides. A parallel formulation for predictive control with nonuniform hold constraints. *Annual Reviews in Control*, 35(2):207–214, 2011a.

S. Longo, E. C. Kerrigan, K. V. Ling, and G. A. Constantinides. Parallel move blocking model predictive control. In *Proceedings of the 50th IEEE Conference on Decision and Control and European Control Conference*, pages 1239–1244, Orlando, FL, USA, 2011b.

S. Lucia, M. Kögel, P. Zometa, D. E. Quevedo, and R. Findeisen. Predictive control, embedded cyberphysical systems and systems of systems–a perspective. *Annual Reviews in Control*, 41:193–207, 2016.

J. MacQueen et al. Some methods for classification and analysis of multivariate observations. In *Proceedings of the 5th Berkeley symposium on mathematical statistics and probability*, volume 1, pages 281–297. California, USA, 1967.

F. Manenti. Considerations on nonlinear model predictive control techniques. *Computers & Chemical Engineering*, 35(11):2491–2509, 2011.

D. Mayne. Robust and stochastic MPC: Are we going in the right direction? *IFAC-PapersOnLine*, 48(23):1–8, 2015.

D. Q. Mayne. Model predictive control: Recent developments and future promise. *Automatica*, 50(12):2967–2986, 2014.

D. Q. Mayne, J. B. Rawlings, C. V. Rao, and P. O. M. Scokaert. Constrained model predictive control: Stability and optimality. *Automatica*, 36(6):789–814, 2000.

J. Mendez, B. Kouvaritakis, and J. Rossiter. State-space approach to interpolation in MPC. *International Journal of Robust and Nonlinear Control*, 10(1):27–38, 2000.

R. Oberdieck and E. N. Pistikopoulos. Explicit hybrid model-predictive control: The exact solution. *Automatica*, 58:152–159, 2015.

G. Pannocchia, J. B. Rawlings, and S. J. Wright. Fast, large-scale model predictive control by partial enumeration. *Automatica*, 43(5):852–860, 2007.

T. Parisini and R. Zoppoli. A receding-horizon regulator for nonlinear systems and a neural approximation. *Automatica*, 31(10):1443–1451, 1995.

S. J. Qin and T. A. Badgwell. A survey of industrial model predictive control technology. *Control engineering practice*, 11(7):733–764, 2003.

D. M. Raimondo, O. Huber, M. S. Darup, M. Mönnigmann, and M. Morari. Constrained time-optimal control for nonlinear systems: a fast explicit approximation. *IFAC Proceedings Volumes*, 45(17):113–118, 2012.

S. Rakovic, A. Teel, D. Mayne, and A. Astolfi. Simple robust control invariant tubes for some classes of nonlinear discrete time systems. In *Proceedings of the 45th IEEE Conference on Decision and Control, 2006*, pages 6397–6402. IEEE, 2006.

J. B. Rawlings and D. Q. Mayne. *Model predictive control: Theory and design*. Nob Hill Pub., 2009.

O. J. Rojas, G. C. Goodwin, M. M. Serón, and A. Feuer. An SVD based strategy for receding horizon control of input constrained linear systems. *International Journal of Robust and Nonlinear Control*, 14(13-14):1207–1226, 2004.

O. Romanko. *Parametric and multiobjective optimization with applications in finance*. PhD thesis, McMaster University, Hamilton, Ontario (Canada), 2010.

J. Rossiter, B. Kouvaritakis, and M. Bacic. Interpolation based computationally efficient predictive control. *International Journal of Control*, 77(3):290–301, 2004.

J. A. Rossiter and P. Grieder. Using interpolation to improve efficiency of multiparametric predictive control. *Automatica*, 41(4):637–643, 2005.

J. A. Rossiter, B. Kouvaritakis, and M. J. Rice. A numerically robust state-space approach to stable-predictive control strategies. *Automatica*, 34(1):65–73, 1998.

J.-H. Ryu, V. Dua, and E. N. Pistikopoulos. A bilevel programming framework for enterprise-wide process networks under uncertainty. *Computers & Chemical Engineering*, 28(6): 1121–1129, 2004.

D. Schlipf, D. J. Schlipf, and M. Kühn. Nonlinear model predictive control of wind turbines using LIDAR. *Wind Energy*, 16(7):1107–1129, 2013.

H. Schmidt. Implementation of a semi-explicit model predictive controller on a two-wheeled mobile robot. Master's thesis, Institute for Systems Theory and Automatic Control (IST), University of Stuttgart, Germany, 2016.

M. Schulze Darup and M. Cannon. Some observations on the activity of terminal constraints in linear MPC. In *Proceedings of the 2016 European Control Conference*, pages 770–775. IEEE, 2016.

P. O. M. Scokaert, D. Q. Mayne, and J. B. Rawlings. Suboptimal model predictive control (feasibility implies stability). *IEEE Transactions on Automatic Control*, 44(3):648–654, 1999.

R. C. Shekhar and C. Manzie. Optimal move blocking strategies for model predictive control. *Automatica*, 61:27–34, 2015.

S. Summers, D. M. Raimondo, C. N. Jones, J. Lygeros, and M. Morari. Fast explicit nonlinear model predictive control via multiresolution function approximation with guaranteed stability. *IFAC Proceedings Volumes*, 43(14):533–538, 2010.

S. Summers, C. N. Jones, J. Lygeros, and M. Morari. A multiresolution approximation method for fast explicit model predictive control. *IEEE Transactions on Automatic Control*, 56(11):2530–2541, 2011.

P. Tøndel, T. A. Johansen, and A. Bemporad. An algorithm for multi-parametric quadratic programming and explicit MPC solutions. *Automatica*, 39(3):489–497, 2003.

J. Unger, M. Kozek, and S. Jakubek. Reduced order optimization for model predictive control using principal control moves. *Journal of Process Control*, 22(1):272–279, 2011.

G. Valencia-Palomo and J. A. Rossiter. Efficient suboptimal parametric solutions to predictive control for PLC applications. *Control Engineering Practice*, 19(7):732–743, 2011.

S. Vazquez, J. I. Leon, L. G. Franquelo, J. Rodriguez, H. A. Young, A. Marquez, and P. Zanchetta. Model predictive control: A review of its applications in power electronics. *IEEE Industrial Electronics Magazine*, 8(1):16–31, 2014.

R. Vidal. Subspace clustering. *IEEE Signal Processing Magazine*, 28(2):52–68, 2011.

Y. Wang and S. Boyd. Fast model predictive control using online optimization. *IEEE Transactions on Control Systems Technology*, 18(2):267–278, 2010.

L. Würth, R. Hannemann, and W. Marquardt. Neighboring-extremal updates for nonlinear model-predictive control and dynamic real-time optimization. *Journal of Process Control*, 19(8):1277–1288, 2009.

Y. Yamamoto. NXTway-GS model-based design-control of selfbalancing two-wheeled robot built with LEGO mindstorms NXT. Technical report, Drexel University, 2008.

M. N. Zeilinger, C. N. Jones, and M. Morari. Real-time suboptimal model predictive control using a combination of explicit MPC and online optimization. *IEEE Transactions on Automatic Control*, 56(7):1524–1534, 2011.

Y. Zhou and C. J. Spanos. On a class of multi-parametric quadratic programming and its applications to machine learning. In *Proceedings of the 55th IEEE Conference on Decision and Control, 2016*, pages 2826–2833. IEEE, 2016.

Q. Zhu, S. Onori, and R. Prucka. Pattern recognition technique based active set QP strategy applied to MPC for a driving cycle test. In *Proceedings of the 2015 American Control Conference*, pages 4935–4940. IEEE, 2015.

P. Zometa, M. Kogel, T. Faulwasser, and R. Findeisen. Implementation aspects of model predictive control for embedded systems. In *Proceedings of 2012 American Control Conference*, pages 1205–1210, Montreal, Canada, 2012.